GRADUATE RECORD EXAMINATION ENGINEERING

Joseph Cataldo, Ph.D.,
Professor, School of Engineering
The Cooper Union School of Art and Science

Yusuf Efe, Ph.D., *Professor, School of Engineering*
The Cooper Union School of Art and Science

Chor Weng Tan, Ph.D.,
Professor, School of Engineering
The Cooper Union School of Art and Science

Alan Wolf, Ph.D., *Assistant Professor, School of Engineering*
The Cooper Union School of Art and Science

ARCO

New York

First Edition

Copyright © 1988 by Joseph Cataldo, Yusuf Efe,
Chor Weng Tan, Alan Wolf
All rights reserved
including the right of reproduction
in whole or in part in any form

ARCO

Simon & Schuster, Inc.
Gulf+Western Building
One Gulf+Western Plaza
New York, NY 10023

DISTRIBUTED BY PRENTICE HALL TRADE

Manufactured in the United States of America

1 2 3 4 5 6 7 8 9 10

Library of Congress Cataloging-in-Publication Data

Graduate record examination in engineering.

  "An Arco book."
  1. Engineering—Examinations, questions, etc.
2. Graduate record examination—Study guides.   I. Cataldo,
Joseph.
TA159.G62   1987   620'.0076   87-17567
ISBN 0-12-363573-2

GRADUATE RECORD EXAMINATION ENGINEERING

CONTENTS

PART ONE

Studying for Your GRE Engineering Test

Using this Book 3

    The Graduate Record Examinations Program/
    Educational Testing Service—The Subject Tests 3

    Content of the GRE Engineering Test 4

    Preparing to Take the Test 5

    Taking the Test 6

    Your Test Score 7

    The Graduate Record Examination General Test 7

    Receiving the Results 8

PART TWO

Two Sample Tests

Sample Test 1

    Answer Sheet 13

    Questions 15

    Answer Key 57

    Solutions 58

Sample Test 2

    Answer Sheet 107

    Questions 109

    Answer Key 154

    Solutions 155

GRADUATE RECORD EXAMINATION ENGINEERING

Part One

STUDYING FOR YOUR GRE ENGINEERING TEST

Using This Book

There are two complete sample Graduate Record Examination (GRE) Engineering Tests in this book. They have a total of 280 multiple-choice problems with their complete solutions. Although the 280 problems are not actual GRE Engineering Test problems found in previous examinations, they are similar in subject matter and approximately the same level of difficulty. Both tests have approximately the same proportion of problems in calculus, mechanics, thermodynamics, and so on, as in actual GRE Engineering Tests.

In Part One of this book, there is a short discussion of the Graduate Record Examinations Program/Educational Testing Service followed by an outline of the content of the GRE Engineering Test and strategies for taking and preparing for the examination.

This book is not a complete study guide to the GRE Engineering Test. Although the solutions are presented in detail, a complete theoretical development of the problems' subject matter is not presented. Therefore, you should supplement your preparation by reviewing basic principles in mathematics, science, and engineering.

THE GRADUATE RECORD EXAMINATIONS PROGRAM/ EDUCATIONAL TESTING SERVICE— THE SUBJECT TESTS

The object of the Graduate Record Examinations Program is to help graduate school admissions committees and granting agencies determine the qualification of applicants in their subject fields. The GRE Tests also help students determine their strengths and weaknesses in their chosen field of study. The tests are standardized so a comparison can be made of different students from colleges throughout the country. There are 20 fields in which the Graduate Record Examinations Program offers a Subject Test:

Biology	Computer Science
Chemistry	Economics

Education	Mathematics
Engineering	Music
French	Philosophy
Geography	Physics
Geology	Political Science
German	Psychology
History	Sociology
Literature in English	Spanish

The Educational Testing Service (ETS) under policies determined by the Graduate Record Examinations Board, an independent board affiliated with the Association of Graduate Schools and the Council of Graduate Schools in the United States, administers the tests.

Booklets describing each test and including sample questions and score interpretation information are available free of charge for all 20 subject tests by writing to

Graduate Record Examinations Program/Educational Testing Service
CN 6000
Princeton, NJ 08541-6000

A full-length GRE Engineering Test given in April 1983 may also be purchased from the Graduate Record Examinations Program by writing to the above address. The answers to this examination are given; however, the solutions are not presented.

CONTENT OF THE GRE ENGINEERING TEST

The GRE Engineering Test consists of approximately 140 multiple-choice questions. There are five choices to each question with one correct or best answer. Some of the questions are based on tables, graphs, and figures; these questions are generally set up in groups.

There are two subscores provided in the test:

1. The engineering subscore, which is based on topics covered in an engineering undergraduate program in most colleges

2. The mathematical subscore, which gauges your ability to answer questions by recalling facts and intuitively approaching problems using calculus

There are approximately 90 engineering questions based on the following areas:

- Mechanics: statics, dynamics, kinematics
- Thermodynamics
- Fluid mechanics and hydraulics
- Transfer and rate mechanics: heat, mass, momentum
- Electricity
- Chemistry
- Nature and properties of matter, including the particulate
- Light and sound
- Computer fundamentals
- Properties of engineering materials
- Engineering economy
- Engineering judgment

There are about 50 questions in the mathematics section of the test. It is assumed that you have taken the following mathematics courses or studied the following topics in your undergraduate curriculum:

- Two courses in calculus
- Differential equations
- Linear algebra
- Numerical analysis
- Probability and statistics

PREPARING TO TAKE THE TEST

You are probably about to complete your undergraduate engineering degree if you are preparing for the GRE Engineering Test. Because of the large range of subject matter covered in this test, it is unlikely that you will be able to answer all the problems. Therefore, you will have to decide what subject matter to focus on in your studies. A reasonable starting point is to review your undergraduate notes and textbooks, starting with theory and fundamentals, and solve a series of problems in the topics outlined above. After completing the review, take the first Sample Test in this book, and score your results (see the section "Your Test Score" on page 7) to determine what your weak and strong subjects are.

When reviewing your test results, determine the reason you missed the correct answer for each question. Was it a lack of knowledge? Did you misread the question? Was your reasoning incorrect? Knowing why you answered the question incorrectly will help you improve your chances of avoiding similar errors.

You may have answered the question correctly but used a different solution from the one in this book. Study the solution given in this book; understanding another correct analysis to the problem will be a good learning experience.

After reviewing your weak subjects, take the second Sample Test in this book using the same procedures as for the first. It is also advisable to obtain the April 1983 GRE Engineering Test from the Graduate Record

Examinations Program. Consider this a third practice test, and take it using the procedures outlined here.

You are not allowed to bring any calculators, slide rules, books, compasses, rulers, dictionaries, or papers into the examination room. If you are found with any of these items, the test paper and answer sheet will be taken from you. The answer sheet will not be scored, and the incident will be reported to the institutions listed on your registration form.

All work must be done on the test paper since you will not be given any scrap paper to work out your solutions. Therefore, when you take the Sample Tests in this book, do not use any of items listed above. There are a few problems (less than 5 percent) in Sample Tests 1 and 2 that may require the use of a scientific hand-held calculator. Time yourself, and make sure that you stay within the time permitted—170 minutes (2 hours and 50 minutes), or about 1.2 minutes per problem. Make sure you take the practice test in a quiet location, and try to complete the test in one sitting. Use a soft pencil (No. 2) to do the calculations and mark the answers on your answer sheet. Only your answers will be graded.

TAKING THE TEST

Make sure to be on time. The examination will be held at the location, time, and day specified on the Admission Ticket of ETS authorization returned to you approximately 1 month before the date of the test. The General Test is usually given in the morning from 8:00 A.M. to approximately 12:55 P.M. and the Subject Test (Engineering Test) in the afternoon from 2:00 P.M. to approximately 4:50 P.M.. You do not have to take the General Test and Engineering Test (or other Subject Test) on the same day. If you take the Engineering Test on a different day than the General Test, however, you must complete two registration forms. Have at least two means of personal identification with you, such as a driver's license or a signed document. Bring a watch to keep on schedule and at least three No. 2 or HB (soft lead) pencils and an eraser.

The Educational Testing Service supervisors will give you time to read all the instructions and fill in the registration sheet.

All the questions have exactly the same weight. Each correct answer has a raw score of 1, and each incorrect answer has a raw score of minus one-quarter ($-\frac{1}{4}$); no credit is taken off your score if you do not answer a question. Therefore, it does not pay to guess at answers to questions you feel you know little or nothing about. But since it is likely you have some knowledge about the problem you are trying to solve, when in doubt, rule out the obvious wrong answers and guess at the remaining choices. You have a better than one-in-four chance of selecting the correct answer if you rule out two choices.

Since you only have a little more than a minute to solve each problem, do not spend time on questions you are unsure of. Make a mark at the side of the question, and if time permits go back to it. Make sure to check frequently that you are making your answers in the appropriate row. A common error in taking standardized tests is to fill in the wrong row.

YOUR TEST SCORE

There are three scores that are reported from the GRE Engineering Test:

1. Total score based on all the questions
2. Subscore in engineering
3. Subscore in mathematics usage

Again, only the answers on the answer sheet are graded. No scrap work is considered part of your score.

There is a raw score determined as follows:

1. Plus 1 for each correct answer
2. Minus one-quarter for each wrong answer
3. No credit for questions left blank

Therefore, your maximum raw score can be 140, assuming there are 140 questions on the test. This raw score is converted into a scaled score, with a minimum of 200 and a maximum of 900. If you leave every question blank, your scaled score will be approximately 340. Remember, no credit is taken off for a question that is left blank. Therefore, if all the questions are blank the raw score will be 0.

Included as part of the registration fee is a booklet with sample problems and information to help you understand what your GRE score means.

THE GRADUATE RECORD EXAMINATION GENERAL TEST

Besides the GRE Subject Tests, such as the Engineering Test, many graduate schools require the GRE General Test. This test is made up of verbal, quantitative, and analytical ability sections. The General Test consists of seven 30-minute sections for a total of $3\frac{1}{2}$ hours. Advanced mathematics is not required for the quantitative ability sections, only a general understanding of algebra and geometry. As in the Engineering Test, there are five multiple choices for each question. Unlike the Engineering Test, however, no credit is taken from your score for wrong answers. Therefore, do not leave any questions blank; guess when you are not sure of an answer.

Although the general discussion in this book about studying and taking the GRE Engineering Test can be applied to the General Test, this book has no questions or specific information on the test. A free bulletin on the General Test may be obtained from the Graduate Record Examinations Program/Educational Testing Service by writing to the address on page 4 or telephoning (609) 771-7670. The "GRE Information Bulletin" contains information on the following:

- Registration
- Preparing for the tests
- Minority Graduate Student Locater Service
- Publications available

For a fee, recent GRE General Tests are also available from the Graduate Record Examinations Program by writing to them at the address on page 4 or calling them at the telephone number mentioned above.

RECEIVING THE RESULTS

As part of the testing fee, the Educational Testing Service will send your scores to you and three schools you designate on your application. It generally takes 4–6 weeks after the test administered to receive the scores of your tests. ETS will send your GRE Test scores to extra schools for an additional fee.

If for any reason after taking the GRE Test you feel your performance was below your capability, you can cancel your score and retake the test at a later date. There is a special form to be filled out for the cancellation of your GRE Test scores.

The scores reported for the Engineering Test are a total score (based on all the questions in the test) and two subscores:

1. Engineering
2. Mathematics usage

For the General Test, the GRE report scores will contain verbal, quantitative, and analytical ability scores.

The ETS has found that there is usually a slight improvement in your score when you retake the General Test. Similar improvement can occur by retaking the Subject Tests, particularly if you intensify your preparation for the second examination. We believe it is best to be adequately prepared for the GRE Tests the first time. If you feel that your score is well below your ability, then consider retaking the test. Note that it is possible that the score on the second test will be less than that on the first test. In that case, both scores will be reported to the graduate schools indicated on your registration form.

The Engineering Test score does not have a percentile ranking. The General Test score does include a percentile ranking, that is, your score relative to others taking the test at the same time. The approximate maximum total score for the Engineering Test is 900; the maximum score for the General Test is 800 for each of the three measures for a total maximum of 2400.

Your three most recent GRE Engineering Test and General Test scores will automatically be reported if you have taken these tests several times.

By filling out the proper request form, you may have earlier test scores (more than the latest three) sent to the institutions of your choice.

The GRE Tests that you take will only be released to the schools designated by you in writing. ETS will not release your scores to graduate schools or agencies without your approval.

Part Two

TWO SAMPLE TESTS

Sample Test 1
Answer Sheet

1 Ⓐ Ⓑ Ⓒ Ⓓ Ⓔ	29 Ⓐ Ⓑ Ⓒ Ⓓ Ⓔ	57 Ⓐ Ⓑ Ⓒ Ⓓ Ⓔ	85 Ⓐ Ⓑ Ⓒ Ⓓ Ⓔ	113 Ⓐ Ⓑ Ⓒ Ⓓ Ⓔ
2 Ⓐ Ⓑ Ⓒ Ⓓ Ⓔ	30 Ⓐ Ⓑ Ⓒ Ⓓ Ⓔ	58 Ⓐ Ⓑ Ⓒ Ⓓ Ⓔ	86 Ⓐ Ⓑ Ⓒ Ⓓ Ⓔ	114 Ⓐ Ⓑ Ⓒ Ⓓ Ⓔ
3 Ⓐ Ⓑ Ⓒ Ⓓ Ⓔ	31 Ⓐ Ⓑ Ⓒ Ⓓ Ⓔ	59 Ⓐ Ⓑ Ⓒ Ⓓ Ⓔ	87 Ⓐ Ⓑ Ⓒ Ⓓ Ⓔ	115 Ⓐ Ⓑ Ⓒ Ⓓ Ⓔ
4 Ⓐ Ⓑ Ⓒ Ⓓ Ⓔ	32 Ⓐ Ⓑ Ⓒ Ⓓ Ⓔ	60 Ⓐ Ⓑ Ⓒ Ⓓ Ⓔ	88 Ⓐ Ⓑ Ⓒ Ⓓ Ⓔ	116 Ⓐ Ⓑ Ⓒ Ⓓ Ⓔ
5 Ⓐ Ⓑ Ⓒ Ⓓ Ⓔ	33 Ⓐ Ⓑ Ⓒ Ⓓ Ⓔ	61 Ⓐ Ⓑ Ⓒ Ⓓ Ⓔ	89 Ⓐ Ⓑ Ⓒ Ⓓ Ⓔ	117 Ⓐ Ⓑ Ⓒ Ⓓ Ⓔ
6 Ⓐ Ⓑ Ⓒ Ⓓ Ⓔ	34 Ⓐ Ⓑ Ⓒ Ⓓ Ⓔ	62 Ⓐ Ⓑ Ⓒ Ⓓ Ⓔ	90 Ⓐ Ⓑ Ⓒ Ⓓ Ⓔ	118 Ⓐ Ⓑ Ⓒ Ⓓ Ⓔ
7 Ⓐ Ⓑ Ⓒ Ⓓ Ⓔ	35 Ⓐ Ⓑ Ⓒ Ⓓ Ⓔ	63 Ⓐ Ⓑ Ⓒ Ⓓ Ⓔ	91 Ⓐ Ⓑ Ⓒ Ⓓ Ⓔ	119 Ⓐ Ⓑ Ⓒ Ⓓ Ⓔ
8 Ⓐ Ⓑ Ⓒ Ⓓ Ⓔ	36 Ⓐ Ⓑ Ⓒ Ⓓ Ⓔ	64 Ⓐ Ⓑ Ⓒ Ⓓ Ⓔ	92 Ⓐ Ⓑ Ⓒ Ⓓ Ⓔ	120 Ⓐ Ⓑ Ⓒ Ⓓ Ⓔ
9 Ⓐ Ⓑ Ⓒ Ⓓ Ⓔ	37 Ⓐ Ⓑ Ⓒ Ⓓ Ⓔ	65 Ⓐ Ⓑ Ⓒ Ⓓ Ⓔ	93 Ⓐ Ⓑ Ⓒ Ⓓ Ⓔ	121 Ⓐ Ⓑ Ⓒ Ⓓ Ⓔ
10 Ⓐ Ⓑ Ⓒ Ⓓ Ⓔ	38 Ⓐ Ⓑ Ⓒ Ⓓ Ⓔ	66 Ⓐ Ⓑ Ⓒ Ⓓ Ⓔ	94 Ⓐ Ⓑ Ⓒ Ⓓ Ⓔ	122 Ⓐ Ⓑ Ⓒ Ⓓ Ⓔ
11 Ⓐ Ⓑ Ⓒ Ⓓ Ⓔ	39 Ⓐ Ⓑ Ⓒ Ⓓ Ⓔ	67 Ⓐ Ⓑ Ⓒ Ⓓ Ⓔ	95 Ⓐ Ⓑ Ⓒ Ⓓ Ⓔ	123 Ⓐ Ⓑ Ⓒ Ⓓ Ⓔ
12 Ⓐ Ⓑ Ⓒ Ⓓ Ⓔ	40 Ⓐ Ⓑ Ⓒ Ⓓ Ⓔ	68 Ⓐ Ⓑ Ⓒ Ⓓ Ⓔ	96 Ⓐ Ⓑ Ⓒ Ⓓ Ⓔ	124 Ⓐ Ⓑ Ⓒ Ⓓ Ⓔ
13 Ⓐ Ⓑ Ⓒ Ⓓ Ⓔ	41 Ⓐ Ⓑ Ⓒ Ⓓ Ⓔ	69 Ⓐ Ⓑ Ⓒ Ⓓ Ⓔ	97 Ⓐ Ⓑ Ⓒ Ⓓ Ⓔ	125 Ⓐ Ⓑ Ⓒ Ⓓ Ⓔ
14 Ⓐ Ⓑ Ⓒ Ⓓ Ⓔ	42 Ⓐ Ⓑ Ⓒ Ⓓ Ⓔ	70 Ⓐ Ⓑ Ⓒ Ⓓ Ⓔ	98 Ⓐ Ⓑ Ⓒ Ⓓ Ⓔ	126 Ⓐ Ⓑ Ⓒ Ⓓ Ⓔ
15 Ⓐ Ⓑ Ⓒ Ⓓ Ⓔ	43 Ⓐ Ⓑ Ⓒ Ⓓ Ⓔ	71 Ⓐ Ⓑ Ⓒ Ⓓ Ⓔ	99 Ⓐ Ⓑ Ⓒ Ⓓ Ⓔ	127 Ⓐ Ⓑ Ⓒ Ⓓ Ⓔ
16 Ⓐ Ⓑ Ⓒ Ⓓ Ⓔ	44 Ⓐ Ⓑ Ⓒ Ⓓ Ⓔ	72 Ⓐ Ⓑ Ⓒ Ⓓ Ⓔ	100 Ⓐ Ⓑ Ⓒ Ⓓ Ⓔ	128 Ⓐ Ⓑ Ⓒ Ⓓ Ⓔ
17 Ⓐ Ⓑ Ⓒ Ⓓ Ⓔ	45 Ⓐ Ⓑ Ⓒ Ⓓ Ⓔ	73 Ⓐ Ⓑ Ⓒ Ⓓ Ⓔ	101 Ⓐ Ⓑ Ⓒ Ⓓ Ⓔ	129 Ⓐ Ⓑ Ⓒ Ⓓ Ⓔ
18 Ⓐ Ⓑ Ⓒ Ⓓ Ⓔ	46 Ⓐ Ⓑ Ⓒ Ⓓ Ⓔ	74 Ⓐ Ⓑ Ⓒ Ⓓ Ⓔ	102 Ⓐ Ⓑ Ⓒ Ⓓ Ⓔ	130 Ⓐ Ⓑ Ⓒ Ⓓ Ⓔ
19 Ⓐ Ⓑ Ⓒ Ⓓ Ⓔ	47 Ⓐ Ⓑ Ⓒ Ⓓ Ⓔ	75 Ⓐ Ⓑ Ⓒ Ⓓ Ⓔ	103 Ⓐ Ⓑ Ⓒ Ⓓ Ⓔ	131 Ⓐ Ⓑ Ⓒ Ⓓ Ⓔ
20 Ⓐ Ⓑ Ⓒ Ⓓ Ⓔ	48 Ⓐ Ⓑ Ⓒ Ⓓ Ⓔ	76 Ⓐ Ⓑ Ⓒ Ⓓ Ⓔ	104 Ⓐ Ⓑ Ⓒ Ⓓ Ⓔ	132 Ⓐ Ⓑ Ⓒ Ⓓ Ⓔ
21 Ⓐ Ⓑ Ⓒ Ⓓ Ⓔ	49 Ⓐ Ⓑ Ⓒ Ⓓ Ⓔ	77 Ⓐ Ⓑ Ⓒ Ⓓ Ⓔ	105 Ⓐ Ⓑ Ⓒ Ⓓ Ⓔ	133 Ⓐ Ⓑ Ⓒ Ⓓ Ⓔ
22 Ⓐ Ⓑ Ⓒ Ⓓ Ⓔ	50 Ⓐ Ⓑ Ⓒ Ⓓ Ⓔ	78 Ⓐ Ⓑ Ⓒ Ⓓ Ⓔ	106 Ⓐ Ⓑ Ⓒ Ⓓ Ⓔ	134 Ⓐ Ⓑ Ⓒ Ⓓ Ⓔ
23 Ⓐ Ⓑ Ⓒ Ⓓ Ⓔ	51 Ⓐ Ⓑ Ⓒ Ⓓ Ⓔ	79 Ⓐ Ⓑ Ⓒ Ⓓ Ⓔ	107 Ⓐ Ⓑ Ⓒ Ⓓ Ⓔ	135 Ⓐ Ⓑ Ⓒ Ⓓ Ⓔ
24 Ⓐ Ⓑ Ⓒ Ⓓ Ⓔ	52 Ⓐ Ⓑ Ⓒ Ⓓ Ⓔ	80 Ⓐ Ⓑ Ⓒ Ⓓ Ⓔ	108 Ⓐ Ⓑ Ⓒ Ⓓ Ⓔ	136 Ⓐ Ⓑ Ⓒ Ⓓ Ⓔ
25 Ⓐ Ⓑ Ⓒ Ⓓ Ⓔ	53 Ⓐ Ⓑ Ⓒ Ⓓ Ⓔ	81 Ⓐ Ⓑ Ⓒ Ⓓ Ⓔ	109 Ⓐ Ⓑ Ⓒ Ⓓ Ⓔ	137 Ⓐ Ⓑ Ⓒ Ⓓ Ⓔ
26 Ⓐ Ⓑ Ⓒ Ⓓ Ⓔ	54 Ⓐ Ⓑ Ⓒ Ⓓ Ⓔ	82 Ⓐ Ⓑ Ⓒ Ⓓ Ⓔ	110 Ⓐ Ⓑ Ⓒ Ⓓ Ⓔ	138 Ⓐ Ⓑ Ⓒ Ⓓ Ⓔ
27 Ⓐ Ⓑ Ⓒ Ⓓ Ⓔ	55 Ⓐ Ⓑ Ⓒ Ⓓ Ⓔ	83 Ⓐ Ⓑ Ⓒ Ⓓ Ⓔ	111 Ⓐ Ⓑ Ⓒ Ⓓ Ⓔ	139 Ⓐ Ⓑ Ⓒ Ⓓ Ⓔ
28 Ⓐ Ⓑ Ⓒ Ⓓ Ⓔ	56 Ⓐ Ⓑ Ⓒ Ⓓ Ⓔ	84 Ⓐ Ⓑ Ⓒ Ⓓ Ⓔ	112 Ⓐ Ⓑ Ⓒ Ⓓ Ⓔ	140 Ⓐ Ⓑ Ⓒ Ⓓ Ⓔ

Sample Test 1

Time—170 Minutes
140 Questions

PART A

Directions: Each of the problems or incomplete statements below is followed by five suggested answers or completions. Select the one that is best in each case and then blacken the corresponding space on the answer sheet.

1. Given the bridge circuit below, determine the value of the resistance R_x in such a way that the bridge is balanced.

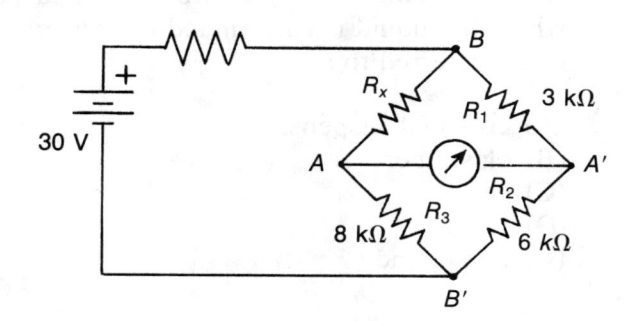

    (A) 2.25 kΩ
    (B) 3 kΩ
    (C) 4 kΩ
    (D) 5 kΩ
    (E) 16 kΩ

2. Which thermodynamic cycle gives the maximum efficiency that can be achieved by an engine operating between two reservoirs at given temperatures?

    (A) Otto cycle
    (B) Diesel cycle
    (C) Carnot cycle
    (D) Rankine cycle
    (E) None of the above

3. A particle of mass m and positive charge q is in a region of constant electric field $\vec{E} = E\hat{z}$ and constant gravitational field $\vec{g} = -g\hat{z}$ ($g = 10$ m/sec^2). How strong is the electric field if the particle has an acceleration of -2 m/sec^2, $\hat{z}$?

   (A) $10m - 2q$
   (B) $\dfrac{10m}{gq}$
   (C) $\dfrac{8m}{q}$
   (D) $\dfrac{12m}{q}$
   (E) none of the above

4. A rectangular slab has a surface area of 1 m^2 and is 4 cm thick. It transmits 10 cal/(sec·cm^2) due to a temperature difference between its end faces. If the coefficient of thermal conductivity is 0.8 cal/(sec·°C·cm) and the cooler end is at room temperature (20°C), what is the temperature of the other face?

   (A) 30°C
   (B) 50°C
   (C) 60°C
   (D) 70°C
   (E) 80°C

5. Asphalt is an ordinary mixture of solid particles and a liquid in which the liquid is surrounded by the solid. Asphalt is most accurately referred to as a

   (A) colloidal suspension
   (B) elastomer
   (C) gel
   (D) glass
   (E) compound

6. A small amount of chalk dust is added to a large quantity of water. How are the freezing point, f, and the boiling point, b, of the water changed?

   (A) b increases, f increases
   (B) b increases, f decreases
   (C) b decreases, f increases
   (D) b decreases, f decreases
   (E) there is no effect on b or f

7. Diving equipment has been designed to withstand an absolute pressure of 5 standard atmospheres in water. The depth a diver can safely descend with this equipment expressed in terms of atmospheric pressures and γ, the specific density of the water, is

(A) 5 atm

(B) $\dfrac{3 \text{ atm}}{\gamma}$

(C) $\dfrac{4 \text{ atm}}{\gamma}$

(D) $\dfrac{5 \text{ atm}}{\gamma}$

(E) $\dfrac{6 \text{ atm}}{\gamma}$

8. In the circuit below, the resistance R is adjusted so that maximum power is delivered to R. Determine the maximum power delivered to R.

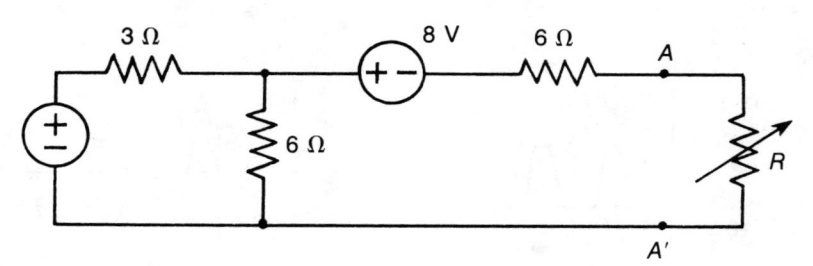

(A) 80 W
(B) 6.25 W
(C) 4.0 W
(D) 3.125 W
(E) none of the above

9. A car of mass m coasts to a stop (on a level road) from speed v in a distance d. Assume all of the friction is between the tires and the road. What is the coefficient of friction μ?

(A) $\mu = \dfrac{gv}{d}$

(B) $\mu = \dfrac{mgd}{v}$

(C) $\mu = \dfrac{mv^2}{2gd}$

(D) $\mu = \dfrac{v^2}{2gd}$

(E) none of the above

10. The diode D in the circuit below is an ideal diode.

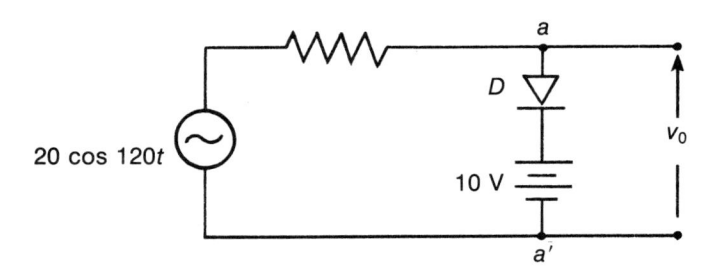

Which of the following would best represent the output voltage, v_0?

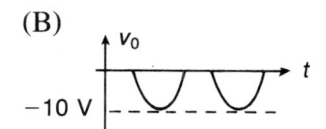

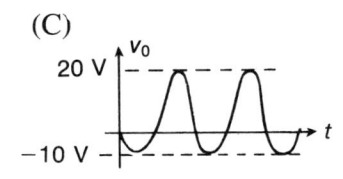

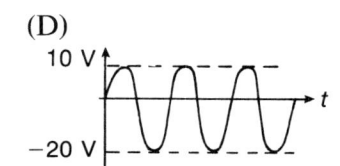

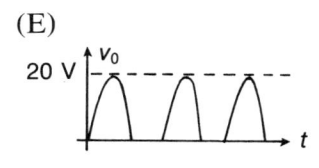

(A)

(B)

(C)

(D)

(E)

11. A cube-shaped piece of metal has linear thermal expansion coefficient α. The volume coefficient of expansion β is most nearly

(A) $\dfrac{\alpha}{3}$

(B) 3α

(C) $\dfrac{\alpha^2}{3}$

(D) α^3

(E) none of the above because β is unrelated to α

12. A train moving directly toward a stationary receiver at 33 m/sec emits a 1200-Hz tone. What frequency is detected? (The speed of sound in air is about 330 m/sec.)

    (A) 540 Hz
    (B) 870 Hz
    (C) 1080 Hz
    (D) 1200 Hz
    (E) 1320 Hz

13. A 40-kg ladder leans against a smooth wall as shown below. What is the coefficient of friction between the ladder and the floor if the ladder is on the verge of slipping? ($g = 10$ m/sec^2)

    (A) 0.2
    (B) 0.4
    (C) 0.5
    (D) 0.8
    (E) 1.0

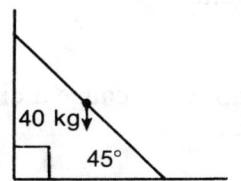

14. The resistance of a copper wire of length l is determined to be R. The wire is cut into four equal pieces, and three of them are placed between two conductive plates as shown. What is the equivalent resistance?

    (A) $12R$
    (B) $3R$
    (C) $\frac{4}{3}R$
    (D) $\frac{1}{3}R$
    (E) $\frac{1}{12}R$

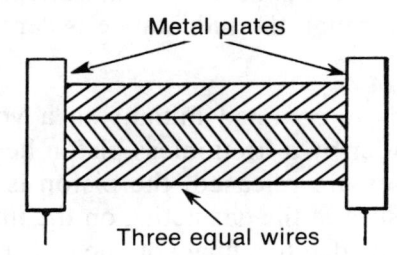

15. A curve whose x values range from -4 to 6 is to be plotted on a computer graphics terminal whose x values must range from -1 to 1. In the terminal coordinates, where will the origin of the curve be found?

    (A) $\frac{2}{3}$
    (B) 0

(C) $\frac{-1}{5}$

(D) $\frac{-1}{4}$

(E) $\frac{-2}{5}$

16. A series circuit is made of a constant voltage source, a capacitor, and a third component. If it is desired to have an oscillating current in the circuit, what should the third component be?

    (A) a resistor
    (B) another capacitor
    (C) an inductor
    (D) a diode
    (E) none of the above because a circuit will not oscillate with a dc voltage source

17. Which of the following statements about inductors is incorrect?

    (A) if there is no change in current through an inductor, the voltage across the inductance is zero
    (B) a finite amount of energy can be stored in an inductor even if the voltage across the inductance is zero
    (C) the current through an inductor cannot be changed in zero time
    (D) an ideal inductor never dissipates energy, but only stores it
    (E) if there is no change in voltage across an inductor, the current through the inductance is zero

18. A gas is confined initially to a volume of 0.5 ft³ in a horizontal cylinder by a frictionless piston held in place by latches. When the latches are released, the piston is forced outward by the internal pressure of the gas acting on the interior piston surface and the gas is expanded to a final volume of 1.0 ft³. A constant external pressure of 25 psi acts on the external piston face. What is the work of the system, including the piston, cylinder, and gas?

    (A) 0.086 ft-lbf
    (B) 12.5 ft-lbf
    (C) 50 ft-lbf
    (D) 1800 ft-lbf
    (E) 3600 ft-lbf

19. Two point particles collide elastically in three dimensions. Conservation laws provide m equations relating the known masses and

initial velocities to the unknown final velocities. To determine the final velocities, n equations are required. What are m and n?

(A) m is 3, n is 3
(B) m is 3, n is 6
(C) m is 4, n is 4
(D) m is 4, n is 6
(E) m is 6, n is 6

20. Consider a piston-and-cylinder apparatus that may be operated so as to bring about the compression or expansion of the gas trapped in the cylinder. The piston and cylinder are assumed to be perfect heat insulators so that no heat can be exchanged between the piston–cylinder apparatus (the system) and its surroundings. Any process carried out in such a system is said to be

(A) isothermal
(B) adiabatic
(C) isentropic
(D) isenthalpic
(E) isobaric

21. Which of the following particles is least useful for a diffraction-based study of atomic structure?

(A) an electron
(B) a proton
(C) an alpha particle
(D) a gamma ray
(E) a neutrino

22. The equation of state of a substance is a

(A) relation between its pressure, specific volume, and temperature
(B) relation between energy, work, and internal energy
(C) thermodynamic state at which the solid, liquid, and gas phases of the substance coexist
(D) relation between its viscosity, thermal conductivity, and diffusivity
(E) statement on the conservation of energy

23. Consider the $n = 2$ to $n = 1$ and the $n = 4$ to $n = 2$ transitions in the hydrogen atom. What is the ratio of the frequencies of radiation that accompany the two transitions?

(A) $1:1$
(B) $1:2$

(C) 1 : 4
(D) 2 : 1
(E) 4 : 1

24. A planet is in an elliptical orbit about a star that sits at one of the foci. At perihelion (the point of closest approach to the star) the orbital velocity is 10^4 m/sec. At aphelion (the point of greatest separation from the star) the orbital velocity is 2.5×10^3 m/sec. The ratio of aphelion distance to perihelion distance is

(A) 1 : 4
(B) 1 : 2
(C) 2 : 1
(D) 4 : 1
(E) cannot be determined from the given information

25. Which of the following set of quantities contains intensive variables?

(A) pressure, temperature, and density
(B) work, energy, and weight
(C) mass, volume, and area
(D) kinetic energy, potential energy, and internal energy
(E) none of the above

26. Three capacitors are connected as shown below.
Determine the electrical charge built up in the capacitor of 3 μF.

(A) 100 μC
(B) 200 μC
(C) 300 μC
(D) 400 μC
(E) 600 μC

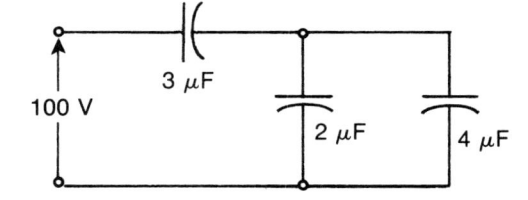

27. Suppose a chemical system at equilibrium is subjected to an increasing temperature. In response, the equilibrium moves in the direction that absorbs heat. This is an example of

(A) conservation of mass–energy
(B) entropy increase in a spontaneous process in a closed system
(C) Le Châtelier's principle
(D) the law of mass–action
(E) maximization of free energy

28. Which of the following happens to a ferromagnet as its temperature is raised above 800°C?

(A) its magnetization increases
(B) its magnetization decreases
(C) its magnetization remains the same
(D) its magnetic domains grow in size
(E) it becomes diamagnetic

29. The frictionless ideal pulley system shown below is used to lift a 100-kg object. What force is required to lift the object at constant speed? (Assume $g = 10$ m/sec.^{2})

    (A) 250 N
    (B) 333 N
    (C) 500 N
    (D) 1000 N
    (E) none of the above

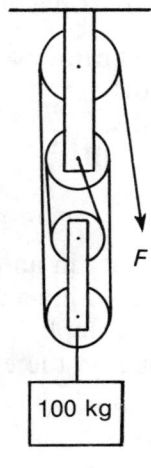

30. What is the empirical formula of a hydrocarbon whose composition is 85.63% carbon and 14.37% hydrogen by weight? (The atomic weight of carbon is 12, of hydrogen approximately 1.)

    (A) CH
    (B) CH_2
    (C) CH_4
    (D) C_2H_6
    (E) C_3H_8

31. An L-shaped beam is fixed in a wall as shown. Find the vertical (V) and horizontal (H) reaction at A.

    (A) $V = 0$ lb
        $H = 0$ lb
    (B) $V = 5.0$ lb
        $H = 8.7$ lb
    (C) $V = 8.7$ lb
        $H = 5.0$ lb
    (D) $V = 5$ lb
        $H = 5$ lb
    (E) $V = 8.7$ lb
        $H = 8.7$ lb

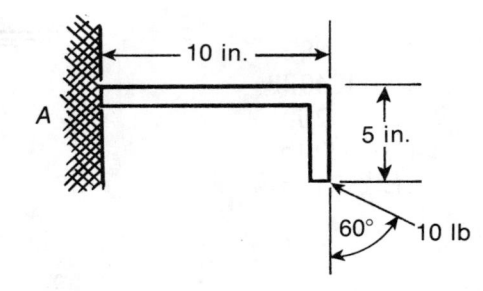

24 / Graduate Record Examination in Engineering

32. Find the moment at A in the figure shown in Problem 31.

    (A) 100 lb·in.
    (B) 86.6 lb·in.
    (C) 61.6 lb·in.
    (D) 50 lb·in.
    (E) 0 lb·in.

33. Three amplifiers are connected in cascade with a gain of 100 dB, 25 dB, and 16 dB each. What is the gain of the overall system?

    (A) 40,000 dB
    (B) 200 dB
    (C) 141 dB
    (D) $\sqrt{141}$ dB
    (E) 19 dB

34. Suppose that in an otherwise empty region of space an electron is brought to within a distance d of a second electron. The work necessary to accomplish this resides in the electric field. The amount of work done on the electron is

$$k = \frac{1}{4\pi\varepsilon_0}$$

    (A) $\dfrac{ke^2}{d^2}$
    (B) $\dfrac{-ke^2}{d^2}$
    (C) $\dfrac{ke^2}{d}$
    (D) $\dfrac{-ke^2}{d}$
    (E) $ke^2 \ln(d)$

35. A wall confines a liquid as shown. A rectangular door 2 × 2 ft rotates freely about hinge A. Find the force F to keep the door in equilibrium. Assume F is acting at the centroid of the hydrostatic force.

    (A) 1488 lb
    (B) 1240 lb
    (C) 992 lb
    (D) 620 lb
    (E) 124 lb

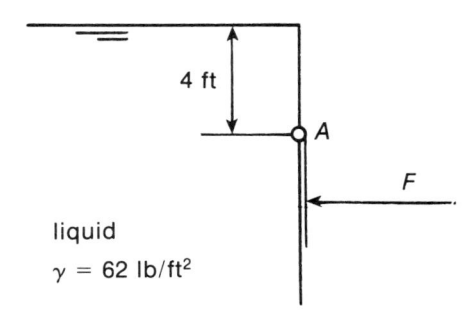

    liquid
    $\gamma = 62$ lb/ft²

36. Consider the chemical reaction A + B → C. The concentrations of these chemicals are a, b, and c, respectively, and the forward reaction constant is k. Initially, $c = 0$. Which of the following equations governs the rate of production of C?

(A) $\dfrac{dc}{dt} = kab$

(B) $\dfrac{dc}{dt} = k(a - c)(b - c)$

(C) $\dfrac{dc}{dt} = \dfrac{kab}{c}$

(D) $\dfrac{dc}{dt} = k(ab - c)$

(E) none of the above

37. The switch in the circuit below has been open for a long time. The current in the circuit at $t = 0^+$ will be

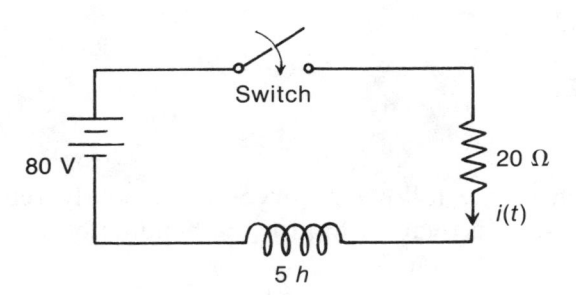

(A) 4 A
(B) 16 A
(C) 3.2 A
(D) 0 A
(E) cannot be determined from the given information

38. A disk-shaped frictionless pulley, $I = 1/2\,(MR^2)$, has a mass of 40 kg and a radius of 400 cm. A rope is wound around the pulley and supports a 4-kg mass. What is the angular acceleration of the pulley? ($g = 10$ m/sec^2)

(A) 0.2 rad/sec^2
(B) 0.4 rad/sec^2
(C) 0.5 rad/sec^2
(D) 1 rad/sec^2
(E) 2 rad/sec^2

39. The wavelength of the fundamental longitudinal vibration for a 2-m pipe that is closed at one end is

   (A) 1 m
   (B) 2 m
   (C) 4 m
   (D) 6 m
   (E) 8 m

40. As the temperature increases in a liquid the viscosity

   (A) increases
   (B) decreases
   (C) remains constant
   (D) increases and then becomes constant
   (E) at first remains constant and then increases

41. A compound made of species A, B, and C has weight ratios of A:B = 2:1 and A:C = 3:2. If the mass of the species are 2, 3, and 4, respectively, what is the smallest possible mass of the molecule?

   (A) 9
   (B) 13
   (C) 18
   (D) 21
   (E) 32

42. Which of the following curves most nearly represents the actual stress–strain relationship for the bending of a steel beam?

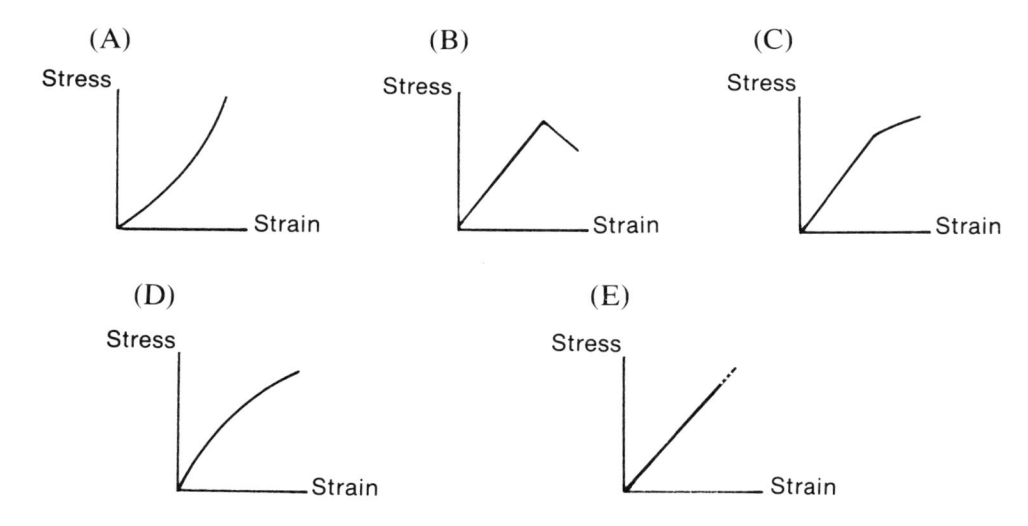

43. An object is placed 1.5 m to the left of a converging thin lens whose focal length f is 50 cm. Where is the image?

    (A) $\frac{1}{2}f$ to the left of the lens

    (B) $\frac{1}{2}f$ to the right of the lens

    (C) $\frac{2}{3}f$ to the right of the lens

    (D) $\frac{3}{2}f$ to the right of the lens

    (E) none of the above

44. The Compton effect describes the scattering of a photon by a massive particle. The quantum mechanical problem differs significantly from the classical problem of electromagnetic radiation scattered by an object in that the

    (A) radiation is absorbed
    (B) radiation always scatters at the same angle
    (C) scattered radiation has a larger wavelength
    (D) scattered radiation has a smaller wavelength
    (E) massive object recoils

45. How long will it take for $20,000 invested at 11% per annum to double in value?

    (A) 6 years

    (B) $6\frac{1}{2}$ years

    (C) 7 years

    (D) $\frac{1}{2}$ year

    (E) 3 years

**End of Part A.
Go to Part B.**

PART B

Directions: For each problem, select the best of the choices offered. Computation and scratch work may be done in this examination book.

46. Determine the following integral:

    $$\int \sin^6 2x \cos 2x \, dx$$

    (A) $\sin^7 2x + c$

    (B) $\frac{1}{14} \sin^7 2x$

    (C) $\frac{1}{14} \sin^7 2x + c$

    (D) $\cos^7 2x + c$

    (E) $\frac{1}{14} \sin^7 2x \cos x$

47. Determine $\frac{dy}{dx}$ for the following:

    $$y = (2x)^2 + 3\sqrt{2x} - \frac{1}{2x}; x > 0$$

    (A) $8x + \frac{3}{2} \sqrt{2} \frac{1}{\sqrt{x}} + \frac{1}{2} \frac{1}{x^2}$

    (B) $16x + \frac{3}{2} \sqrt{2} \frac{1}{\sqrt{x}} + \frac{\ln x}{2}$

    (C) $16x + \frac{3}{2} \sqrt{2} \frac{1}{\sqrt{x}} + \frac{1}{2} \frac{1}{x^2}$

    (D) $8x - \frac{1}{\sqrt{2x}} + \frac{1}{2} \frac{1}{x^2}$

    (E) $8x + \frac{3}{2} \sqrt{2} \frac{1}{\sqrt{x}}$

28

48. Find the directed length of the projection of $\vec{V} = 3\vec{i} + 2\vec{j}$ on the direction of $\vec{T} = -6\vec{i} + 2\vec{j}$.

   (A) $\dfrac{-14}{\sqrt{40}}$

   (B) $\dfrac{14}{\sqrt{40}}$

   (C) 14

   (D) $-6\vec{i} + 2\vec{j}$

   (E) $\dfrac{22}{\sqrt{40}}$

49. $\displaystyle\int_0^1 (2x + 4x^3 - 3)\, dx =$

   (A) -1
   (B) 0
   (C) $+1$
   (D) $+3$
   (E) $+5$

50. The matrix product

$$\begin{vmatrix} 4 & 2 \\ 1 & 0 \end{vmatrix} \begin{vmatrix} 1 & 3 & 0 \\ 2 & 0 & 1 \end{vmatrix}$$

   is

   (A) $\begin{vmatrix} 8 & 3 & 0 \\ 12 & 2 & 0 \end{vmatrix}$

   (B) $\begin{vmatrix} 8 & 12 & 2 \\ 1 & 3 & 0 \end{vmatrix}$

   (C) $\begin{vmatrix} 2 & 12 & 8 \\ 0 & 4 & 1 \end{vmatrix}$

   (D) $\begin{vmatrix} 1 & 3 & 0 \\ 2 & 0 & 1 \end{vmatrix}$

   (E) $\begin{vmatrix} 1 & 2 \\ 3 & 0 \\ 0 & 1 \end{vmatrix}$

51. Which of the following matrices is an identity matrix?

   (A) $\begin{vmatrix} 1 & 0 \\ 1 & 0 \end{vmatrix}$

(B) $\begin{vmatrix} 1 & 0 \\ 0 & 1 \end{vmatrix}$

(C) $\begin{vmatrix} 2 & 0 \\ 0 & 2 \end{vmatrix}$

(D) $\begin{vmatrix} 1 & 1 \\ 1 & 1 \end{vmatrix}$

(E) $\begin{vmatrix} 1-i & 0 \\ 0 & 1+i \end{vmatrix}$

52. $(1 + x)(1 - x + x^2 - x^3 + x^4 - x^5 + x^6) =$

    (A) x^7
    (B) $1 + x^7$
    (C) $1 - x^7$
    (D) $1 + x^6$
    (E) $1 - x^6$

53. If

$$A = \begin{pmatrix} 3 & 2 \\ 1 & 4 \\ 1 & 1 \end{pmatrix}$$

and

$$B = \begin{pmatrix} 3 & 1 & 1 \\ 2 & 4 & -2 \end{pmatrix}$$

then

$(AB)^T =$

(A) $(11 \quad 18)$

(B) $\begin{pmatrix} 11 & 3 \\ 12 & 4 \end{pmatrix}$

(C) $\begin{pmatrix} 13 & 11 & 5 \\ 11 & 17 & 5 \\ -1 & -7 & -1 \end{pmatrix}$

(D) $\begin{pmatrix} 11 & 4 \\ 13 & -9 \end{pmatrix}$

(E) $\begin{pmatrix} 13 & 11 & -1 \\ 11 & 17 & -7 \\ 5 & 5 & -1 \end{pmatrix}$

54. Matrix A is

$$\begin{pmatrix} 3 & 2 \\ 1 & -1 \end{pmatrix}$$

Which of the following matrices *cannot* be obtained from matrix A by the operations of Gaussian elimination?

(A) $\begin{pmatrix} 3 & 2 \\ 0 & 0 \end{pmatrix}$

(B) $\begin{pmatrix} 3 & 2 \\ 4 & 1 \end{pmatrix}$

(C) $\begin{pmatrix} 2 & 3 \\ 1 & -1 \end{pmatrix}$

(D) $\begin{pmatrix} 6 & 4 \\ 2 & -2 \end{pmatrix}$

(E) $\begin{pmatrix} 6 & 4 \\ 1 & -1 \end{pmatrix}$

55. Consider the following system of equations:

$$2x + 4y = 9$$
$$3x - 3y = 11$$

Using Cramer's rule to solve this system, the determinant of the "denominator matrix" is

(A) 18
(B) 9
(C) 0
(D) −2
(E) −18

56. A pair of numbers is randomly picked (without replacement) from the set {1, 2, 3, 5, 7, 11, 12, 13, 17, 19}. What is the probability that the number 11 was picked given that the sum of the numbers was even?

(A) 0.1
(B) 0.125
(C) 0.18
(D) 0.24
(E) 0.38

57. In Problem 56, what is the probability that the number 11 was picked (at least once) if the choices were made with replacement?

(A) 0.05
(B) 0.125
(C) 0.18
(D) 0.22
(E) 0.28

58. A solution of the pair of differential equations

$$\dot{x} = y$$
$$\dot{y} = -ky - px$$

where k and p are nonzero constants is of the form

(A) $x = e^{(a+ib)t}$
(B) $x = \sin(\omega t)$
(C) x is a polynomial in t
(D) $x = \sinh(\omega t)$
(E) none of the above

59. Using a two-term Taylor series expansion for the integrand, evaluate

$$\int_0^{0.1} e^{-x^2}\, dx$$

approximately.

(A) 0.0655
(B) 0.0997
(C) 0.1240
(D) 0.2420
(E) 0.9900

60. $\int_y^{1/y} \left(x - \dfrac{1}{x^2}\right) dx =$

(A) $y - \dfrac{1}{y^2}$

(B) $\dfrac{y^2}{2} + \dfrac{1}{y}$

(C) $\dfrac{1}{2y^2} - \dfrac{y}{2}$

(D) $\frac{1}{2y^2} + y - \frac{y^2}{2} - \frac{1}{y}$

(E) none of the above

61. $\int_0^{2\pi} \sin^2(x)\,dx =$

(A) 2π

(B) π

(C) $\frac{\pi}{2}$

(D) 1

(E) 0

62. With the appropriate change of variable of integration and integration limits

$$\int \frac{1}{1+2x^2}\,dx$$

can be written as

(A) $\int \sec^2\theta\,d\theta$

(B) $\frac{1}{\sqrt{2}}\int d\theta$

(C) $\frac{1}{\sqrt{2}}\tan^{-1}(x)$

(D) $\int \frac{\sec(x)}{\tan(x)}\,dx$

(E) $\int \frac{4x}{1+2x^2}\,dx$

63. $\int \tan(x)\,dx =$

(A) $\tan(x)\cos^2(x)$
(B) $-\ln[\cos(x)]$
(C) $\ln[\sin(x)]$
(D) $\cot(x)$
(E) none of the above

64. The curve $y = x^3$ from $x = 0$ to $x = 2$ is revolved about the y axis to form a cuplike object in three dimensions. The volume enclosed by this object (up to $y = 8$) is

(A) $\int_0^8 \frac{4}{3} \pi y^3 \, dy$

(B) $\int_0^8 \pi y^{2/3} \, dy$

(C) $\int_0^8 2\pi y^{1/3} \, dy$

(D) $\int_0^2 \pi x^6 \, dx$

(E) $\int_0^2 2\pi x^3 \, dx$

**End of Part B.
Go to Part C.**

PART C

Directions: For each problem, select the best of the choices offered. Every function whose graph appears in this part of the test is to be assumed to have derivatives of all orders at each point of its domain unless otherwise indicated.

Problems 65–70 are based on the following information.

Functions f and g have derivatives of all orders.

65. Given

   $$\int_a^b f(x)\, dx$$

   what integral results from the transformation

   $$x = g(t)$$

   (A) $\int_\alpha^\beta f[g(t)]\, dt$, for suitable α, β

   (B) $\int_\alpha^\beta f[g(t)]g'(t)\, dt$, for suitable α, β

   (C) $\int_a^b f[g(t)]g'(t)\, dt$

   (D) $\int_a^b f(t)g'(t)\, dt$

   (E) $\int_a^b f[g(t)]\, dt$

66. The integral

   $$\int_a^b f'(x)g'(x)\, dx$$

   is equal to

(A) $f(x)g(x)\Big|_a^b$

(B) $\int_a^b f'(x)g(x)\,dx + \int_a^b f(x)g'(x)\,dx$

(C) $f'(x)g(x)\Big|_a^b + \int_a^b f'(x)g(x)\,dx$

(D) $f'(x)g(x)\Big|_a^b - \int_a^b f''(x)g(x)\,dx$

(E) none of the above

67. If $f(x)$ is a linear function of x and $g(x)$ is a quadratic function of x, the integrand $1/fg$ may be replaced by which of the following simpler integrands (the k's are constants)?

(A) $\dfrac{1}{k_1 x + k_2}$

(B) $\dfrac{k_1 f}{g} + \dfrac{k_2 g}{f}$

(C) $\dfrac{k_1}{f} + \dfrac{k_2}{g}$

(D) $k_1 f + k_2 g$

(E) $\dfrac{k_1}{f} + \dfrac{k_2 x + k_3}{g}$

68. $x' = f(x, y)$ and $y' = g(x, y)$ are well-behaved coordinate transformations. What is the ratio of infinitesimal area elements $\dfrac{dx'dy'}{dxdy}$ at $x = x_0$, $y = y_0$?

(A) $\dfrac{\partial f}{\partial x}\Big|_{x_0} \dfrac{\partial g}{\partial y}\Big|_{y_0}$

(B) $\dfrac{\partial f}{\partial y}\Big|_{(x_0,y_0)} \dfrac{\partial g}{\partial x}\Big|_{(x_0,y_0)}$

(C) $\dfrac{\partial^2 f}{\partial x \partial y}\Big|_{(x_0,y_0)} \dfrac{\partial^2 g}{\partial x \partial y}\Big|_{(x_0,y_0)}$

(D) $\left(\dfrac{\partial f}{\partial x}\dfrac{\partial g}{\partial y} - \dfrac{\partial f}{\partial y}\dfrac{\partial g}{\partial x}\right)\Big|_{(x_0,y_0)}$

(E) $\left(\dfrac{\partial f}{\partial x}\dfrac{\partial g}{\partial y} + \dfrac{\partial f}{\partial y}\dfrac{\partial g}{\partial x}\right)\Big|_{(x_0,y_0)}$

69. $\dfrac{d}{dx} g\left(\dfrac{1}{x}\right)$ for $g(x) = x^2 + \dfrac{1}{x}$ is

   (A) $2x + 1$

   (B) $2x - \dfrac{1}{x^2}$

   (C) $1 - \dfrac{2}{x^3}$

   (D) $\dfrac{-2}{x^2}$

   (E) $\dfrac{2}{x} - x^2$

70. If $f(x)$ is everywhere decreasing and $g'(x)$ is always negative, then f/g is

   (A) positive, but not necessarily a monotone function
   (B) positive and monotone
   (C) negative, but not necessarily a monotone function
   (D) negative and monotone
   (E) not necessarily any of the above

Problems 71–75 are based on the following information.

The function $f(t)$ plotted below is infinitely differentiable. The function F is defined by

$$F(x) = \int_0^x f(t)\, dt \quad \text{for } 0 \leq t \leq 15$$

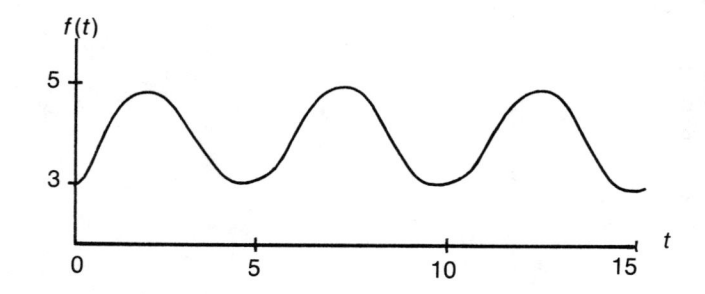

71. On the interval shown, $F(x)$ is

   (A) strictly increasing
   (B) strictly decreasing
   (C) periodic
   (D) alternately increasing and decreasing
   (E) none of the above

72. On the interval shown, the number of times $f''(t) = 0$ is

 (A) 0
 (B) 1
 (C) 3
 (D) 4
 (E) 6

73. Of the following, the best approximation to $F(5)$ is

 (A) 0
 (B) 3
 (C) 5
 (D) 10
 (E) 20

74. The minimum value of the derivative of F on the interval (0, 15) is

 (A) −1
 (B) 0
 (C) 1
 (D) 3
 (E) 5

75. The best approximation of the average value of the derivative on the interval (0, 15) is

 (A) 0
 (B) 3
 (C) 4
 (D) 5
 (E) 8

Problems 76–80 are based on the following information.

An object X begins at a vertex A of a nonregular hexagon and moves along the perimeter to vertices B, C, D, E, F and then back to A. There is a 1-sec pause at each vertex, otherwise the speed is constant. The graphs on the following page show the straight-line distance of X from points B and E as a function of time. The trip is completed in 20 sec.

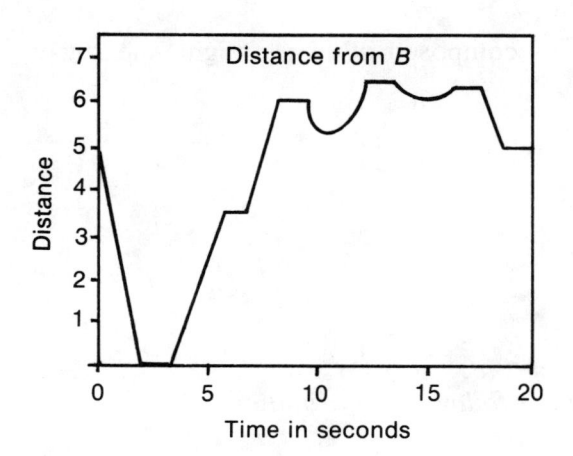

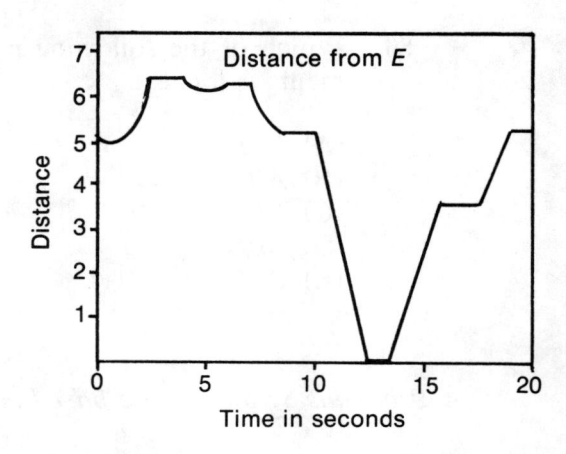

76. The distance from A to B is about

   (A) 2
   (B) 5
   (C) 7
   (D) 10
   (E) 11

77. The straight-line distance from B to E is about

   (A) 4
   (B) 5
   (C) 6.4
   (D) 8
   (E) 9.5

78. Which vertex is most nearly equidistant from vertices B and E?

   (A) A
   (B) C
   (C) D
   (D) F
   (E) two vertices are tied

79. The triangle with vertices B, D, and E is

   (A) a right triangle
   (B) isosceles but not equilateral
   (C) equilateral
   (D) one that contains an angle between 160° and 180°
   (E) none of the above

80. Which of the following paths composed of two straight line segments is longest?

    (A) FED
    (B) ABC
    (C) EDB
    (D) ABD
    (E) BEC

Problems 81 and 82 are based on the following information.

The figure below is a probability density function (p.d.f.).

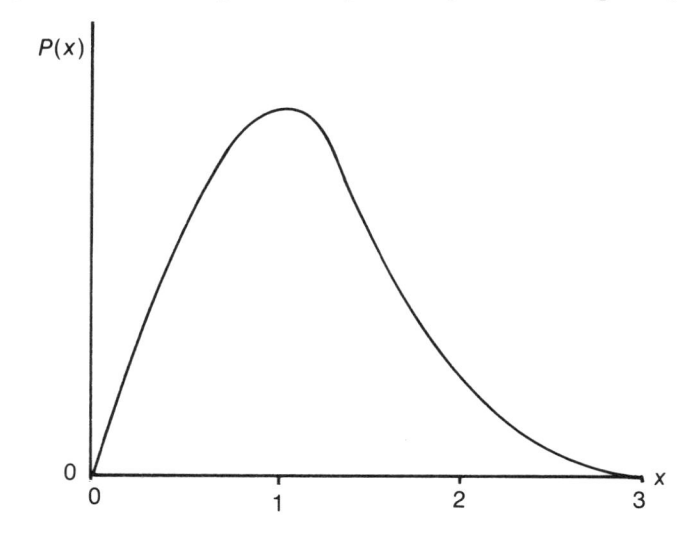

81. Which of the following graphs is the cumulative density function that belongs to the above p.d.f.?

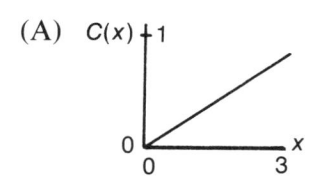

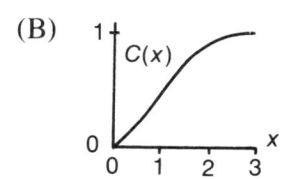

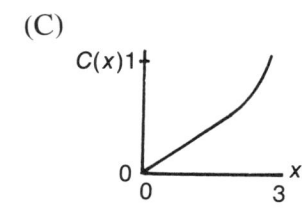

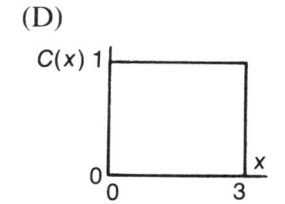

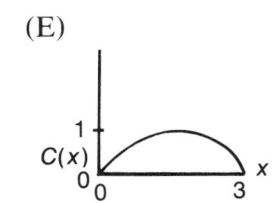

    (A)
    (B)
    (C)
    (D)
    (E)

82. Which of the following values of x corresponds to a change in concavity of the probability distribution function?

    (A) 0
    (B) 0.8
    (C) 1.3
    (D) 2
    (E) 3

83. $\dfrac{d}{dx}[\tan(x)\sin(x)] =$

    (A) $\tan(x)\cos(x) + \sec^2(x)$
    (B) $\cos(x)$
    (C) $\ln[\tan(x)\sec(x)]$
    (D) $\tan^2(x)\cos(x)$
    (E) none of the above

84. $\dfrac{d}{dx}(2)^{x^2} =$

    (A) x^2
    (B) 2^{x^2}
    (C) $2^{x^2}\ln 2$
    (D) $2^{x^2}(\ln 2)(2x)$
    (E) $2x(2^{x^2})$

85. If money earning 5% compounded annually is compounded monthly instead, what is the new effective annual interest rate?

    (A) 5%
    (B) $(0.05)^{12} \times 100\%$
    (C) $\left(\left(1 + \dfrac{0.05}{12}\right)^{12} - 1\right) \times 100\%$
    (D) $\left(\dfrac{0.05}{12}\right)^{12} \times 100\%$
    (E) $((1.05)^{12} - 1.05) \times 100\%$

End of Part C.
Go to Part D.

PART D

Directions: Each of the problems or incomplete statements below is followed by five suggested answers or completions. Select the one that is best in each case and then blacken the corresponding space on the answer sheet.

86. Two objects of mass *m* traveling at half the speed of light in opposite directions collide inelastically. A negligible amount of their total energy leaves the masses as sound, light, and so on. The single composite object that is formed

    (A) has a mass of 2*m*
    (B) has a mass larger than 2*m*
    (C) has a mass smaller than 2*m*
    (D) travels at half the speed of light to conserve momentum
    (E) has no net momentum in any inertial frame

87. Two capacitors of 3 μF and 6 μF are connected in series across a 30-V power supply. The charge stored in the capacitor of 6 μF is

    (A) 60 μC
    (B) 30 μC
    (C) 10 μC
    (D) 5 μC
    (E) none of the above

88. Consider the chemical reaction

    $$H_2(g) + Cl_2(g) = 2HCl(g)$$

    The thermodynamic equilibrium constant K_p for the reaction is given by:
    (the partial pressures are P_i, the chemical . . .)

    (A) $K_p = \dfrac{P_{H_2} P_{Cl_2}}{P_{HCl}^2}$

    (B) $K_p = \dfrac{P_{HCl}^2}{P_{H_2} P_{Cl_2}}$

    (C) $K_p = \dfrac{C_{HCl}}{C_{H_2} C_{Cl_2}}$

    (D) $K_p = C_{H_2} C_{Cl_2} C_{HCl}^2$

    (E) $K_p = C_{HCl}^2 + C_{H_2} C_{Cl_2}$

89. Determine the value of the reactance of the capacitor in the circuit below such that the current in the circuit is maximum.

    (A) $(40 + j30)\Omega$
    (B) $(40 - j30)\Omega$
    (C) $-30\ \Omega$
    (D) $30\ \Omega$
    (E) $60\pi\ \Omega$

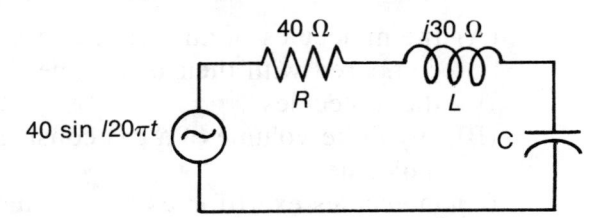

90. Given a container whose cross section is shown in the figure below; h equals a depth of water. If the surface areas at AA', BB', CC', DD' are a, $3a$, $2a$, and a, respectively, which surface is subjected to the largest force? (Neglect the pressure of atmosphere.)

    (A) AA'
    (B) BB'
    (C) CC'
    (D) DD'
    (E) both (C) and (B)

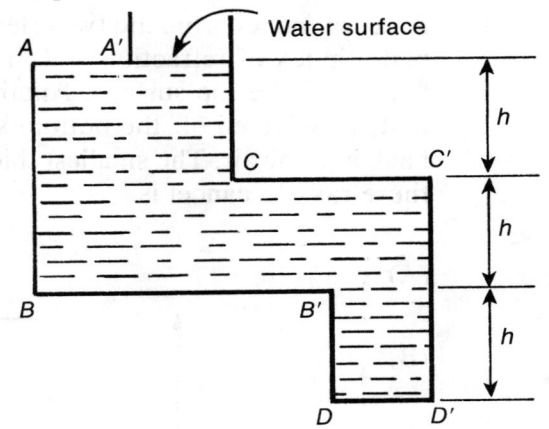

91. A coil of 240 turns is cut by a magnetic field that decreases at the rate of 50 Wb/min. If the resistance of the coil is 40 Ω and its inductance is 2 H, what is the current through the coil?

    (A) 0 A
    (B) 1.25 A
    (C) 4 A
    (D) 5 A
    (E) 6 A

92. The specific weight of the liquid in the container whose cross section is shown below is γ and the container is W units wide. Find the magnitude of the upward force on the inclined wall.

    (A) $\gamma\sqrt{2}\,hW$
    (B) $\dfrac{\gamma h^2 W}{\sqrt{2}}$
    (C) $\dfrac{\gamma h^2 W}{2}$
    (D) γhW
    (E) γh

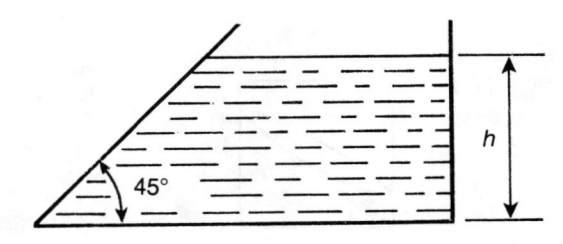

93. Which of the following statements describes one of the basic assumptions of an ideal gas?

    (A) the molecules of an ideal gas are separated by distances large compared with their own dimensions, that is, the diameter of the molecules
    (B) any finite volume of a gas consists of a small, finite number of molecules
    (C) molecules exert forces on one another
    (D) molecules collide with one another inelastically
    (E) the molecules are distributed normally throughout the container

94. A ray of monochromatic (wavelength = λ) light striking a thin glass plate (index of refraction = 1.5) at almost normal incidence is reflected off the top surface. Another ray, initially in phase with the first, is reflected off the bottom surface and ultimately transmitted back into the air. The smallest thickness t (other than 0) required for these rays to cancel is

    (A) $\frac{\lambda}{3}$

    (B) $\frac{\lambda}{4}$

    (C) $\frac{\lambda}{6}$

    (D) λ

    (E) not dependent on λ

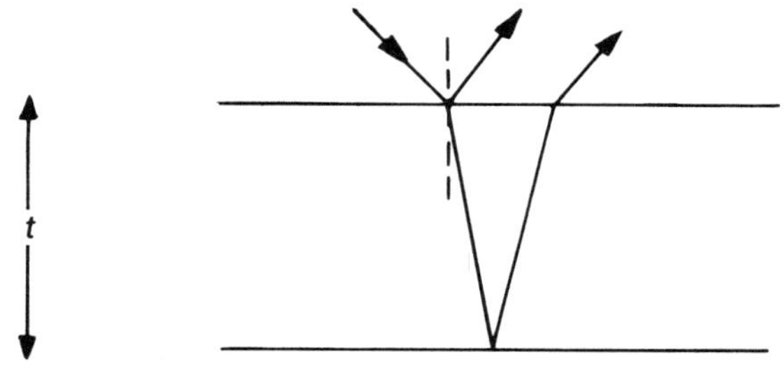

95. For an ideal gas, the absolute temperature T is proportional to

    (A) mean square speed of the molecules of the gas
    (B) root-mean-square velocity of the gas molecules
    (C) average velocity of the gas molecules
    (D) pressure and volume of the gas
    (E) pressure of the gas only

96. The phase diagram for water (at moderate pressures) looks most like

    (A) P vs T

    (B) P vs T

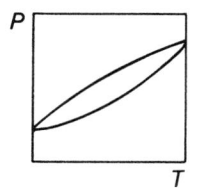

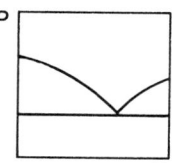

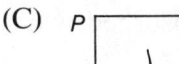

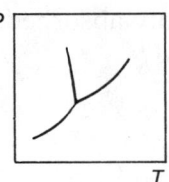

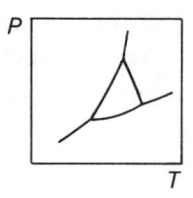

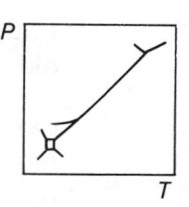

(C) ![P vs T graph]
(D) ![P vs T graph]
(E) ![P vs T graph]

97. Unpolarized light passes through two polarizers whose polarizing axes have a relative rotation of $\theta°$. If the intensity of the emerging light is one-sixth that of the incident light, what is θ?

    (A) $0°$

    (B) $\cos^{-1}\left(\dfrac{1}{\sqrt{2}}\right)$

    (C) $\cos^{-1}\left(\dfrac{1}{\sqrt{3}}\right)$

    (D) $\cos^{-1}\left(\dfrac{1}{3}\right)$

    (E) $\cos^{-1}\left(\dfrac{1}{2}\right)$

98. A fluid traveling through a pipe of varying cross section A at velocity $\vec{v}$ is observed to satisfy the equation $A|\vec{v}| = $ constant. This is most likely because the

    (A) pressure is constant throughout the fluid
    (B) density is constant throughout the fluid
    (C) fluid is Newtonian
    (D) fluid travels at constant velocity
    (E) pressure is proportional to the density

99. A hollow cylindrical object rolls down an inclined plane without slipping. Which of the following factors influences its velocity at the bottom?

    (A) radius of object
    (B) length of object
    (C) density of object
    (D) change in height of object during trip
    (E) smoothness of the cylinder's surface

100. Radiocarbon dating is based on the fact that for many thousands of years living organisms have been absorbing a small known amount of radioactive carbon-14, which decays with a half-life of 5730 years. If radioactive carbon can be accurately measured in concen-

trations down to 6% of the amount typically absorbed, what is the oldest sample that can be accurately dated?

(A) 10,000 years
(B) 13,000 years
(C) 18,000 years
(D) 23,000 years
(E) 28,000 years

101. If a particle's mass is doubled and its nonrelativistic velocity is cut in half, its quantum mechanical wavelength will

(A) quadruple
(B) double
(C) remain the same
(D) be halved
(E) be quartered

102. The height of water (h) in the tank whose cross section is shown below remains constant for the given conditions in the figure. For a vertical inflow of 100 ft^3/sec and a horizontal opening on the side of the tank with an area of 2.5 ft^2, the velocity exiting the tank v is most nearly (neglect all loss terms)

(A) 250 ft/sec
(B) 40 ft/sec
(C) 20 ft/sec
(D) 4 ft/sec
(E) 2 ft/sec

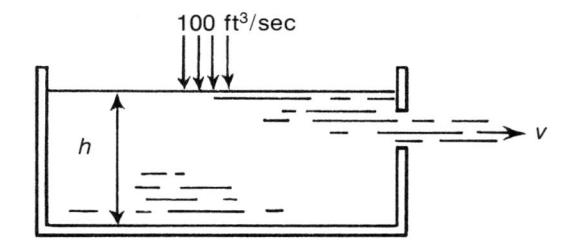

103. An atom is populated by some number of electrons such that the total angular momentum and the total electron spin are zero. Which of the following might be the number of electrons?

(A) 3
(B) 6
(C) 8
(D) 10
(E) 11

104. The sodium chloride crystal structure may be created by putting Na$^+$ and Cl$^-$ ions alternatively at lattice points of a simple cubic lattice. Another way to view this structure is to consider an Na$^+$ and

a Cl⁻ lattice that are interpenetrating. Each of these lattices are of what type?

(A) face-centered cubic
(B) body-centered cubic
(C) hexagonal close packed
(D) simple cubic
(E) a glass

105. The expression

$$dU = C_v dT$$

is valid for

(A) an ideal gas regardless of the process
(B) a real substance at constant pressure
(C) a real substance independent of V and p
(D) all real gases at high pressures
(E) none of the above

106. A piece of machinery cost (first cost) $20,000. It has been estimated that this machine has a life of 10 years and a salvage value of $2,000 at the end of this time. Determine the annual depreciation cost for this machine assuming a straight line method.

(A) $18,000
(B) $2,200
(C) $2,000
(D) $1,800
(E) $1,600

107. Which of the following particles is most affected by the presence of a strong magnetic field at right angles to the particle's velocity? (Assume the particles all travel at the same speed.)

(A) alpha particle
(B) neutrino
(C) gamma particle
(D) neutron
(E) electron

108. The curve of nuclear binding energy (binding energy per nucleon as a function of atomic mass) explains how both fission and fusion processes release energy. The behavior of this curve with (increasing) mass is

(A) monotone up
(B) monotone down
(C) monotone up, then monotone down
(D) monotone down, then monotone up
(E) monotone down, up, then down

109. If you exactly double the number density of an ideal gas in a given vessel, you would make

(A) the pressure twice as great
(B) the pressure four times as great
(C) the speed of every molecule twice as great
(D) the absolute temperature twice as great
(E) the absolute temperature four times as great

110. A given manometer is filled with two homogeneous liquids with specific density $\bar{\gamma}$ for the lower liquid and γ for the upper liquid, $\bar{\gamma} > \gamma$. The difference in pressure $P_A - P_B$ is

(A) $(2\bar{\gamma} + \gamma)L$
(B) $(\bar{\gamma} + \gamma)L$
(C) $\bar{\gamma}L$
(D) γL
(E) 0

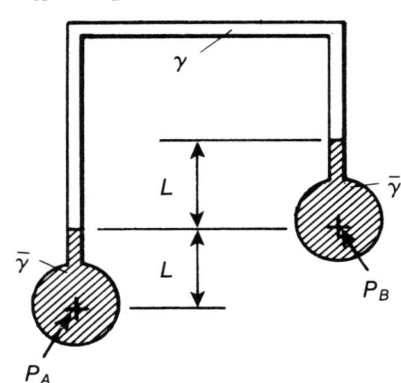

111. For an ideal gas, the ratio of the specific heats,

$$\frac{C_p}{C_v} = -\frac{V}{p}\left(\frac{\partial p}{\partial V}\right)_s$$

(C_p and C_v are specific heats at constant pressure and volume, respectively; V = volume; p = pressure; s = entropy.) Which of the following statements is *not* true?

(A) $\dfrac{C_p}{C_v} > 0$

(B) $\dfrac{C_p}{C_v} < 0$

(C) Since $\left(\dfrac{\partial p}{\partial V}\right)_s < 0$ for an ideal gas, $\dfrac{C_p}{C_v} > 0$

(D) C_p and C_v are always positive

(E) C_p is very nearly equal to C_v for liquid water

112. A cylindrical bucket is filled with water (specific weight γ) to a height h. Find the pressure at the bottom of the bucket if the bucket is allowed to fall freely. Ignore the friction.

(A) γg

(B) γh

(C) $\dfrac{\gamma h}{2}$

(D) $2\gamma h$

(E) 0

113. A 12-kg mass rests on a surface for which the coefficient of friction is $\mu = 0.15$. The smallest force that can give the mass an acceleration of 3 m/sec² is most nearly:

(A) 120 N
(B) 100 N
(C) 54 N
(D) 42 N
(E) 22 N

114. The velocity and acceleration of a mass on a spring are

(A) in phase
(B) 90° out of phase
(C) 180° out of phase
(D) out of phase by a time-varying amount
(E) out of phase and the velocity always lags by less than 90°

115. The rate at which solar energy impinges on the surface of a unit area placed normal to the sun at the outer fringes of the earth's atmosphere is about 1.4×10^3 J/sec·m². If a space vehicle were flying around the earth at two-third's the distance from the sun to the earth, how much solar energy would strike it on the sunlit side?

(A) $6.4 \times 10^3 \dfrac{J}{\text{sec} \cdot \text{m}^2}$

(B) $3.2 \times 10^3 \dfrac{J}{\text{sec} \cdot \text{m}^2}$

(C) $2.1 \times 10^3 \dfrac{J}{\text{sec} \cdot \text{m}^2}$

(D) $0.9 \times 10^3 \dfrac{J}{\text{sec} \cdot \text{m}^2}$

(E) $0.6 \times 10^3 \dfrac{J}{\text{sec} \cdot \text{m}^2}$

116. A 60-kg chunk of material is placed in a tank filled with water. In order to keep it submerged, a force of 200 N down must be applied. What is the volume of the object? (Density of water = 1000 kg/m³, $g = 10$ m/sec.²)

(A) 0.02 m³
(B) 0.05 m³
(C) 0.08 m³
(D) 0.11 m³
(E) 0.14 m³

117. What third force restores equilibrium to the system below?

(A) 20 N, $\vec{W}$
(B) 42 N, $\vec{E}$
(C) 35 N, $\vec{NW}$
(D) 48 N, $\vec{NE}$
(E) 26 N, $\vec{SW}$

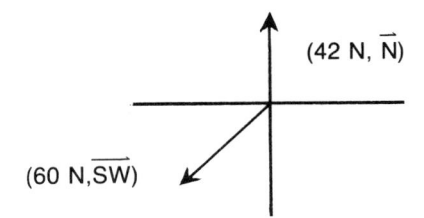

(42 N, $\vec{N}$)

(60 N, $\vec{SW}$)

118. A rectangular open container is partially filled with water and is shown below at rest. If this container is moved horizontally to the right at a constant uniform acceleration, the pressure at *A* will

(A) decrease
(B) increase
(C) stay the same
(D) first decrease then stay the same
(E) oscillate between higher and lower pressure

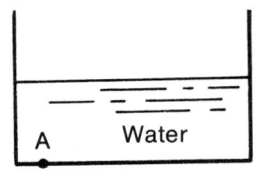

119. A dc generator is rated at 5 V and 4 A at full load and has a voltage regulation of 5%. The Thevenin's equivalent to this generator is

(A) 5 V, 0.0625 Ω
(B) 5 V, 1.25 Ω
(C) 5 V, 1.281 Ω
(D) 5.125 V, 1.281 Ω
(E) 5.25 V, 0.0625 Ω

120. A horse pulls a cart, causing it to move at a constant speed of 2 m/sec. The mass of the horse is 100 kg, that of the cart 200 kg. Friction is negligible except at the horses hooves. What is the ratio of the force of the horse on the cart to that of the cart on the horse?

(A) 1 : 1   (D) 2 : 1
(B) 1 : 2   (E) 20 : 1
(C) 1 : 40

121. In the figure below, what is the reading of the spring scale?

   (A) 0 lb
   (B) 10 lb
   (C) 20 lb
   (D) 40 lb
   (E) none of above

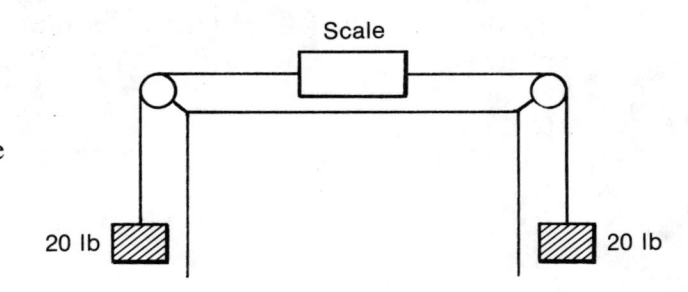

122. The fluid mechanical quantity

   $$\frac{3\pi\eta\mu R^4}{2h^3}$$

   where

   η = viscosity $\left[\dfrac{g}{cm \cdot sec}\right]$

   μ = fluid velocity $\left[\dfrac{cm}{sec}\right]$

   R = radius

   h = height

   has the same units as

   (A) density
   (B) force
   (C) pressure
   (D) vorticity (curl of fluid velocity)
   (E) none of the above

123. What is the work done by the reversible expansion of an ideal gas in a piston-and-cylinder assembly at 300°K from an initial pressure of 10 atm to a final pressure of 1 atm?

   (A) 3000 J/kg-mol
   (B) 1,083,000 J/kg-mol
   (C) 2,494,000 J/kg-mol
   (D) 5,743,000 J/kg-mol
   (E) 24,942,000 J/kg-mol

124. The horizontal force exerted on a fixed vane by a jet of water with a jet area $A = 1$ in.^{2} having a jet velocity of $v = 12$ ft/sec, if the water density $\rho = 2$ slugs/ft^3 and the water is deflected 90° in a horizontal plane is

(A) 288 lb
(B) 144 lb
(C) 24 lb
(D) 2 lb
(E) 0 lb

125. For the horizontal closed pipe shown, the velocity at section 2 is twice the velocity at section 1. If the specific weight of the water expressed in pounds per cubic foot is equal to twice the gravitational constant g, what is the drop in pressure between section 1 and section 2 expressed in terms of the velocity v_1?

(A) v_1^2
(B) $2v_1^2$
(C) $3v_1^2$
(D) $3v_1$
(E) 0

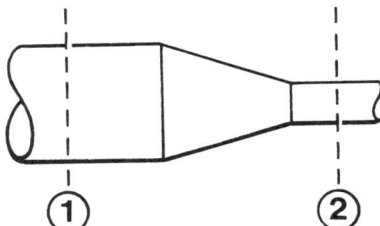

126. For a given reservoir, water is discharging into the atmosphere H feet below the water surface through a relatively small circular opening. The theoretical velocity at the discharge in terms of the head H and gravitational constant g, neglecting all losses, is

(A) 0
(B) $2gH$
(C) $\sqrt{gH}$
(D) $\sqrt{2gH}$
(E) H^2

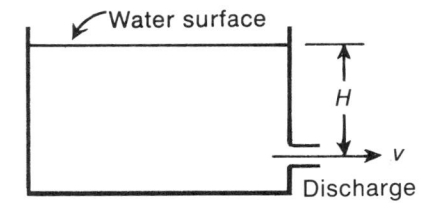

127. For fluid flow in a circular pipe the general velocity vector can be expressed as a function of the time t and coordinates x, y, z, that is, $f(x, y, z, t)$. For a steady one-dimensional flow, the velocity vector may be expressed by which one of the following functional relationships?

(A) $f(t)$
(B) $f(t, x)$
(C) $f(x)$
(D) $f(x, y)$
(E) $f(\text{constant})$

**End of Part D.
Go to Part E.**

PART E

Directions: For each problem select the best of the choices offered. Computation and scratch work may be done in this examination book.

128. For a data set that is normally distributed, how many standard deviations centered about the mean contain about 90% of the data?

    (A) it depends on the variance and mean of the data set
    (B) it depends on the variance of the data set
    (C) less than 1/2 standard deviation
    (D) about 2 standard deviations
    (E) about 10 standard deviations

129. The first term in the Taylor series for tan(x) about 0 is

    (A) 1
    (B) x
    (C) $-x$
    (D) $\frac{x^2}{2}$
    (E) $\frac{-x^2}{2}$

130. $\lim\limits_{n\to\infty} \sum\limits_{0}^{n} \frac{1}{3^k} =$

    (A) $\frac{1}{3}$
    (B) $\frac{2}{3}$
    (C) 1
    (D) $\frac{3}{2}$
    (E) $\frac{4}{3}$

131. The sum of

$$1 + \frac{1}{3} + \frac{1}{2} + \frac{1}{9} + \frac{1}{4} + \frac{1}{27} + \frac{1}{8} + \cdots$$

is

(A) 2

(B) $2\frac{1}{3}$

(C) $2\frac{1}{2}$

(D) e

(E) π

132. Consider the system of equations

$$x + y = 7, \ z - 4x + 5y = 0$$

The graph of the intersection of these equations in three-dimensional space is

(A) a point
(B) a line
(C) a plane
(D) two planes
(E) the null set

133. The intersection of the cone

$$x^2 + y^2 = z$$

with the plane $x = 2$ is a(an)

(A) circle
(B) parabola
(C) plane
(D) hyperbola
(E) ellipse

134. Two temperature scales y and z are linearly related to the old scale x as follows

(1) $\qquad y = 3x + 60$

(2) $\qquad z = x$ at $x = -24$

$\qquad z = -x$ at $x = +8$

Which of the following describes the relationship between scales y and z?

(A) $z = \frac{3}{2}y - 28$

(B) $z = \frac{y}{6} - 22$

(C) $z = \frac{5}{2}y + 48$

(D) $z = \frac{2}{3}y + 72$

(E) none of the above

135. Which of the following best describes the behavior of the function

$$\sin(x)\frac{e^x}{x^x}$$

for large x?

(A) it asymptotes to zero in an oscillatory manner
(B) it asymptotes to zero from above
(C) it diverges
(D) it approaches a finite (nonzero) value
(E) none of the above

136. The expression $\ln(x^y)$ is the same as

(A) $(\ln y)(\ln x)$
(B) $(\ln x)^y$
(C) $y(\ln x)$
(D) $x(\ln y)$
(E) none of the above

137. $e^{i2\pi} =$

(A) 0
(B) 1
(C) -1
(D) $\frac{\pi}{2}$
(E) π

138. The iterative process defined by

$$X_{n+1} = X_n - \frac{(X_n^2 - 5)}{X_n}$$

results in a sequence of numbers that can converge (depending upon X_1) to

(A) 0
(B) $\sqrt{5}$
(C) 5
(D) $-\infty$
(E) ∞

139. The parametric description of the position of an object in three dimensions is

$$x = t^2, \ y = t, \ z = \frac{1}{2} - 2t^2$$

where t is the time. What is the speed of the object as a function of time?

(A) $(2t, 1, -4t)$
(B) $t - t^2$
(C) $2t + 1 - 4t^2$
(D) $\sqrt{2t + 1 - 4t^2}$
(E) $\sqrt{1 + 20t^2}$

140. In 200 experimental measurements whose values ranged from 9 to 23, the mean value was found to be 15 and the variance 9. How many standard deviations was the highest score above the mean?

(A) $\frac{8}{9}$
(B) 2
(C) $\frac{8}{3}$
(D) 3
(E) $\frac{10}{3}$

End of Test 1.

Sample Test 1
Answer Key

1.	C	36.	B	71.	A	106.	D
2.	C	37.	D	72.	E	107.	E
3.	C	38.	C	73.	E	108.	C
4.	D	39.	E	74.	D	109.	A
5.	C	40.	B	75.	C	110.	D
6.	B	41.	B	76.	B	111.	B
7.	C	42.	C	77.	C	112.	E
8.	D	43.	D	78.	A	113.	C
9.	D	44.	C	79.	E	114.	B
10.	D	45.	C	80.	E	115.	B
11.	B	46.	C	81.	B	116.	C
12.	E	47.	A	82.	C	117.	B
13.	C	48.	A	83.	E	118.	B
14.	E	49.	A	84.	D	119.	E
15.	C	50.	B	85.	C	120.	A
16.	C	51.	B	86.	B	121.	C
17.	E	52.	B	87.	A	122.	B
18.	D	53.	C	88.	B	123.	D
19.	D	54.	A	89.	D	124.	D
20.	B	55.	E	90.	B	125.	C
21.	E	56.	D	91.	D	126.	D
22.	A	57.	D	92.	C	127.	C
23.	E	58.	A	93.	A	128.	D
24.	D	59.	B	94.	A	129.	B
25.	A	60.	D	95.	A	130.	D
26.	B	61.	B	96.	C	131.	C
27.	C	62.	B	97.	C	132.	B
28.	B	63.	B	98.	B	133.	B
29.	A	64.	B	99.	D	134.	B
30.	B	65.	B	100.	D	135.	A
31.	C	66.	D	101.	C	136.	C
32.	C	67.	E	102.	B	137.	B
33.	C	68.	D	103.	D	138.	B
34.	C	69.	C	104.	A	139.	E
35.	B	70.	E	105.	A	140.	C

Sample Test 1
Solutions

PART A

1. Correct answer: **(C)**
   The bridge circuit in Problem 1 is usually used for measuring the value of resistances. When the bridge is balanced, that is, when the galvanometer indicates that there is no current flowing through it, the following relationship holds:

   $$R_x \cdot R_2 = R_1 \cdot R_3$$

   Hence,

   $$R_x \cdot 6 = 3 \cdot 8$$

   or

   $$R_x = 4 \text{ k}\Omega$$

2. Correct answer: **(C)**
   An ideal engine is one operating in a particularly simple cycle known as the Carnot cycle. The cycle is characterized as containing two isentropic (reversible adiabatic) and two isothermal processes.

3. Correct answer: **(C)**
   The electric field is the force per unit charge, or $\vec{F} = q\vec{E}$. Similarly, the gravitational force (or weight) is $m\vec{g}$. There are two opposing forces whose sum results in an acceleration of -2 m/sec². From Newton's Second Law,

   $$(qE - mg)\hat{z} = m(-2\hat{z})$$

   or

   $$E = \frac{m(g-2)}{q} = \frac{8m}{q}$$

4. Correct answer: **(D)**
   The thermal flux (heat energy per unit area per unit time) across a slab is equal to the product of the temperature difference between the end faces and the thermal conductivity divided by the thickness

of the slab. The equation

$$\frac{K\Delta T}{l} = 10 \frac{cal}{sec \cdot cm^2} = \frac{0.8 \left(\frac{cal}{sec \cdot °C \cdot cm}\right)(T - 20°C)}{4 \text{ cm}}$$

can be solved for $T = 70°C$. The area of the slab is not needed for this calculation.

5. **Correct answer: (C)**
   Here we use the process of elimination. A colloidal suspension is a primarily liquid mixture containing extremely small solid particles. An elastomer is a substance that can be deformed to twice its natural length or longer. Glass refers to a single homogeneous substance. Asphalt is a mixture of indefinite proportions and therefore is not a compound.

6. **Correct answer: (B)**
   Adding a small quantity of solute to a solvent lowers the freezing point and raises the boiling point of the pure solvent. These effects are related to the lowering of vapor pressure (recall that it's hard to boil water on top of a mountain!) and may be derived from the Clausius–Clapeyron equation. In the case of freezing point depression, the direction of the change should be clear from the everyday example of water and salt.

7. **Correct answer: (C)**
   The equation for the absolute pressure is

   $$P_{abs} = P + \gamma h$$

   where h is the depth the diver can safely descend.
   The maximum pressure the diving equipment can safely be exposed to is

   $$5 \text{ atm} = 1 \text{ atm} + \gamma h$$

   Therefore,

   $$\gamma h = 5 \text{ atm} - 1 \text{ atm} = 4 \text{ atm}$$

   and

   $$h_{max} = \frac{4 \text{ atm}}{\gamma}$$

8. Correct answer: **(D)**
First replace the circuit to the left of AA' with its Thevenin equivalent. You should find the following circuit:

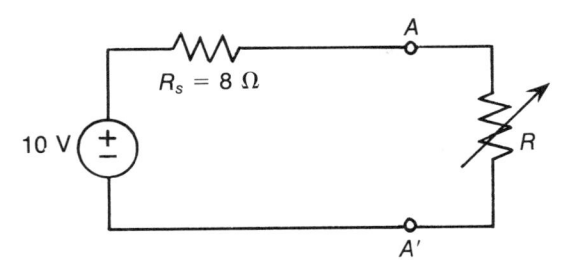

For the maximum power delivered to R, the following condition must be satisfied:

$R_s = R$

Hence,

$R_s = R = 8\ \Omega$

The power delivered to R is

$$P_{max} = i^2 \cdot R = \left(\frac{10}{8+8}\right)^2 \times 8$$

$$P_{max} = 3.125\ \text{W}$$

9. Correct answer: **(D)**
The initial energy of the car is $\frac{1}{2}mv^2$. If this much frictional work is done on the car, it will come to rest. The frictional work is the frictional force, μN, times the distance d. The normal force N is the weight mg.

$$\frac{1}{2}mv^2 = \mu mgd$$

$$\mu = \frac{v^2}{2gd}$$

This problem may also be solved by dimensional analysis.

10. Correct answer: **(D)**
When the diode is conducting, the voltage across the diode D will be zero, hence $v_0 = 10$ V.
When the diode is not conducting, that is, the voltage across aa' is -20 V, $v_0 = -20$ V. Hence, the output voltage varies between $+10$ and -20 V. Therefore, v_0 would best be represented by the graph in (D).

11. Correct answer: **(B)**
The equation for linear thermal expansion of a solid is

$$L = L_0(1 + \alpha[T - T_0]),$$

where L_0 is the length at temperature T_0, L is the length at temperature T, and α is the coefficient of linear expansion. If a cube shows this linear growth in each dimension, the volume becomes $(L_0[1 + \alpha \Delta T])^3$, which is a fractional increase of $(1 + \alpha \Delta T)^3 - 1$. Expanding and ignoring terms $(\alpha \Delta T)^2$ and higher order, we have

$$\Delta V/V = 3\alpha \Delta T = \beta \Delta T,$$

so

$$\beta \approx 3\alpha.$$

12. Correct answer: **(E)**
The equation for the Doppler shift in frequently for a moving source is

$$f' = f\left(\frac{1}{1 \pm \frac{V_1}{V_2}}\right)$$

where f is the source frequency, f' is the measured frequency, V_1 is the speed of the source, and V_2 is the speed of the wave in the medium. The positive sign in the denominator is for relative recession, the negative sign for relative approach. Here

$$f = 1200 \text{ Hz}$$
$$V_1 = 33 \text{ m/sec}$$
$$V_2 = 330 \text{ m/sec}$$

so

$$f' = 1200 \text{ Hz} \left(\frac{1}{1 - \frac{33}{330}}\right) \approx 1200 \text{ Hz} \left(1 + \frac{33}{330}\right)$$

$$= 1200 \text{ Hz } (1.1) = 1320 \text{ Hz}$$

13. Correct answer: **(C)**
The 400-N downward force of the ladder must be counteracted by the vertical component of the force at the floor. The horizontal forces on the ladder are equal and opposite at the two points of

contact. Summing torques around the base of the ladder we have

$$+F_x \frac{l}{\sqrt{2}} - 400 \frac{l}{2\sqrt{2}} = 0$$

or $F_x = 200$ N. This force is frictional in nature and has its maximum value (the ladder is about to slip) of μN. Since N = 400 Newtons, $\mu = 0.5$.

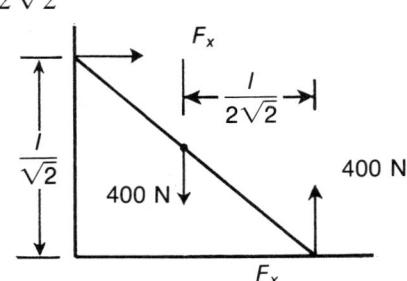

14. Correct answer: **(E)**
The resistance of a wire can be computed by

$$R = \rho \frac{l}{A}$$

where ρ = resistivity, l = length, and A = cross-sectional area of the wire. When the wire is cut into four pieces and three of them placed between two conductive plates, the equivalent resistance, R_{eq}, is given by

$$R_{eq} = \rho \cdot \frac{l/4}{3A} = \frac{1}{12} \cdot \rho \frac{l}{A} = \frac{1}{12} R$$

15. Correct answer: **(C)**
The coefficients of the linear relationship $x' = ax + b$ that takes the interval $(-4, 6)$ into the interval $(-1, 1)$ are found from the simultaneous equations

$$-1 = -4a + b$$
$$1 = 6a + b$$

Solving these, we find $a = 1/5$, $b = -1/5$, and so $x' = (x - 1)/5$. If $x = 0$, the screen coordinate x' is $-1/5$.

16. Correct answer: **(C)**
The voltage drops due to various circuit elements are

$$\text{Resistor} \quad IR = \frac{dQ}{dt} R$$

$$\text{Capacitor} \quad \frac{Q}{C}$$

$$\text{Inductor} \quad \frac{dI}{dt} L = \frac{d^2Q}{dt^2} L$$

dc voltage source ε

If some combination of these elements is wired in series, the sum of their voltage drops must total zero. For example, a battery, resistor, and capacitor give

$$\varepsilon - IR - \frac{Q}{C} = 0$$

If we are to obtain an oscillating solution to such a differential equation, we must have a variable and its second derivative appearing in the form

$$\frac{d^2Q}{dt^2} \propto -Q$$

The only circuit element listed that can provide the second derivative is the inductor.

17. **Correct answer: (E)**
The claims in Problem 17 can be checked by using the following relationship:

$$v(t) = L\frac{di(t)}{dt}$$

We can easily observe that the incorrect statement is in (E).

18. **Correct answer: (D)**
For a constant external pressure P and a volume change of V, the work of the system is

$$W = P\Delta V$$
$$= 25\,\frac{\text{lb}_f}{\text{in.}^2} \cdot 144\,\frac{\text{in.}^2}{\text{ft}^2} \cdot 0.5\text{ ft}^3$$
$$= 1800\text{ ft-lb}_f$$

We have made no use of the pressure of the gas in the cylinder. Had we taken the gas alone as our system, we would have needed the internal pressure exerted on the interior piston face as a function of V.

19. **Correct answer: (D)**
In a perfectly elastic three-dimensional collision between two point masses there are four scalar conservation laws that relate initial and final velocities: one equation for conservation of energy and three equations for conservation of momentum; m is 4. There are a total of six unknowns in the problem: the three components of each of the two final velocities; n is 6. If a detailed interaction mechanism

such as gravity were specified, there would be as many equations as unknowns.

20. Correct answer: **(B)**
By definition, any adiabatic process is one in which no heat transfer can occur between the system and its surroundings.

21. Correct answer: **(E)**
Diffraction is a useful tool in determining atomic structure—a regular array of atoms will scatter an incident beam of particles into a distinctive regular pattern—if the particles have a substantial interaction with the atomic lattice. Electrons, protons, and alpha particles interact electrically with the lattice and are likely to be strongly diffracted (although the heavier the particle, the less the deflection). Gamma rays are uncharged but still interact with atoms via the Compton effect and by inducing atomic transitions. Neutrinos are uncharged and hardly interact with anything and so are not significantly diffracted.

22. Correct answer: **(A)**
The equation of state of a substance describes the relationship of three thermodynamic properties such as temperature, pressure, and specific volume.

23. Correct answer: **(E)**
The energy levels of the hydrogen atom are

$$E_n = \frac{-13.6}{n^2} \text{ eV}$$

In a transition from $n = 2$ to $n = 1$, the energy difference is

$$-13.6 \left(\frac{1}{1^2} - \frac{1}{2^2}\right) \text{ eV} = \frac{3}{4}(-13.6 \text{ eV})$$

and appears as a photon with $E_{21} = h\nu_{21}$, with h being Planck's constant. In the $n = 4$ to $n = 2$ transition, the energy difference E_{42} is

$$-13.6 \text{ eV} \left(\frac{1}{2^2} - \frac{1}{4^2}\right) = \frac{3}{16}(-13.6 \text{ eV})$$

The ratio of photon frequencies is the same as the ratio of photon energies:

$$\frac{\nu_{21}}{\nu_{42}} = \frac{E_{21}}{E_{42}} = \frac{3/4}{3/16} = 4 : 1$$

24. **Correct answer: (D)**
   In the figure below, we see the star at one focus of the elliptical orbit. Perihelion and aphelion correspond to moments when the orbital velocity is at right angles to the vector connecting planet and star. Conservation of orbital angular momentum means

$$|\vec{L}| = |\vec{r} \times \vec{p}| = rp = mvr$$

is a constant. Therefore,

$$mv_p r_p = mv_a r_a$$

or

$$\frac{r_a}{r_p} = \frac{v_p}{v_a} = \frac{10^4}{2.5 \times 10^3} = 4$$

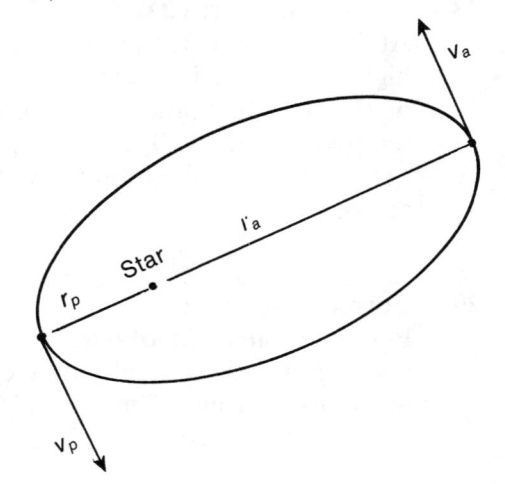

25. **Correct answer: (A)**
   An intensive variable is one whose value is independent of the mass of the system.

26. **Correct answer: (B)**
   The following circuit (on the left) is equivalent to the circuit (on the right):

$$\frac{1}{C} = \frac{1}{C_1} + \frac{1}{C_2} = \frac{1}{3} + \frac{1}{6} = \frac{1}{2} \quad \text{or} \quad C = 2 \ \mu F$$

Charge:

$$Q = C \cdot V = 2 \times 100 = 200 \ \mu C$$

(In a series connection, the charge in each capacitor will be the same.) Hence,

$$Q_{C_1} = Q_{C_2} = Q_C = 200 \ \mu F$$

27. **Correct answer: (C)**
Le Châtelier's principle states that a system in equilibrium responds to any externally imposed stress in the direction that would tend to reduce that stress. In the case of heating such a system, the system will attempt to reduce the thermal stress by absorbing excess heat.

28. **Correct answer: (B)**
At intermediate temperatures, ferromagnets are characterized by high magnetic alignment of atoms in spite of the randomizing effect of thermal fluctuations. When the temperature exceeds the Curie temperature (800°C), however, the ferromagnet becomes paramagnetic—its magnetization then decreases inversely with the temperature.

29. **Correct answer: (A)**
Pulley problems involving a single long rope are easy to solve when you recall that an ideal pulley can only change the direction of the tension in a rope. The 1000-N load is supported by four ropes of equal tensions, so

$$F = \frac{1000 \text{ N}}{4} = 250 \text{ N}$$

If the mass was accelerating up or down, more or less force would be required.

30. **Correct answer: (B)**
The weight ratio of carbon to hydrogen in the hydrocarbon is 6:1, and the weight ratio of a carbon atom to a hydrogen atom is 12:1. Therefore, there must be two hydrogen atoms for each carbon atom, and possible formulas are CH_2 and C_2H_4.

31. **Correct answer: (C)**
Σ Vertical forces

$$V_A = (10 \text{ lb})\sin 60° = 8.7 \text{ lb}$$

Σ Horizontal forces

$$H_A = (10 \text{ lb})\cos 60° = 5.0 \text{ lb}$$

32. **Correct answer: (C)**

$$\Sigma \text{ Moment at } A = (10 \text{ lb})\sin 60°(10 \text{ in.}) - (10 \text{ lb})(\cos 60°)(5 \text{ in.})$$
$$= 86.6 \text{ lb} \cdot \text{in.} - 25 \text{ lb} \cdot \text{in.}$$
$$= 61.6 \text{ lb} \cdot \text{in. a counterclockwise moment}$$

Therefore, the reaction at A is clockwise.

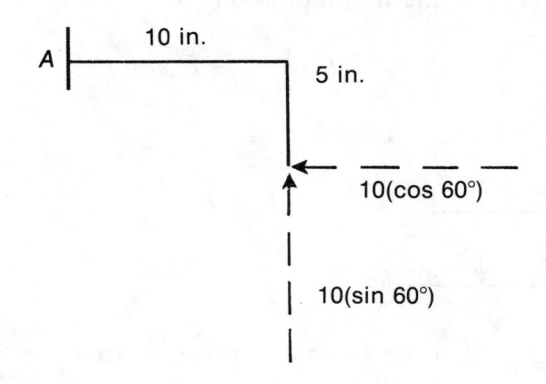

```
A |——— 10 in. ———┐
                 | 5 in.
                 |
              ←————————
              ↑   10(cos 60°)
              |
              | 10(sin 60°)
              |
```

33. **Correct answer: (C)**

    Let the gains of the amplifiers be G_1, G_2, and G_3. Then the overall gain of the cascaded system will be

    $$G = G_1 \cdot G_2 \cdot G_3$$

    In terms of decibels, this can be written as

    $$20 \log G = 20 \log G_1 + 20 \log G_2 + 20 \log G_3$$

    or

    $$G|_{dB} = G_1|_{dB} + G_2|_{dB} + G_3|_{dB}$$

    Hence, overall gain $= 100 + 25 + 16 = 141$ dB.

34. **Correct answer: (C)**

    The work done by a force $\vec{F}(\vec{r})$ through a displacement from $\vec{r}_1$ to $\vec{r}_2$ is given by

    $$\int_{\vec{r}_1}^{\vec{r}_2} \vec{F}(\vec{r}) \cdot d\vec{r}$$

    Here

    $$\vec{F}(\vec{r}) = \frac{ke^2}{r^2} \hat{r}$$

    so the integral is

    $$\int_{\vec{r}_1}^{\vec{r}_2} \vec{F}(\vec{r}) \cdot d\vec{r} = -\int_{\infty}^{d} \frac{ke^2 dr}{r^2} = \left. \frac{ke^2}{r} \right|_{\infty}^{d} = \frac{ke^2}{d}$$

    The work done on the electron is positive because the force must overcome the particles' repulsion.

35. Correct answer: **(B)**
Take the sum of the moment about A

$$F_H \times d = F \times d$$

therefore

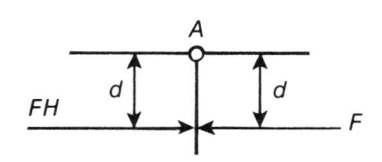

$$F = F_H$$

The hydrostatic force F_H = avg press × area

$$F_H = \frac{4\gamma + 6\gamma}{2} \times 2 \times 2$$

$$= 5(62)4 = 1240 \text{ lb}$$

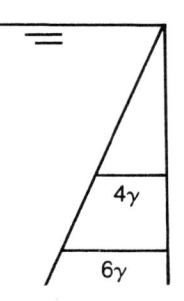

36. Correct answer: **(B)**
The laws of chemical kinetics for bimolecular reactions state that the rate of formation of a product is proportional to the product of the instantaneous concentrations of the reacting species and the rate constant. The instantaneous concentration of A is the amount initially present minus that already consumed, or $(a - c)$. For reacting species B, the instantaneous concentration is $(b - c)$.

37. Correct answer: **(D)**
Since the switch has been open for a long time, $i(t) = 0$ for $t < 0$. The current in an inductor cannot be changed instantaneously. Hence, the current in the circuit at $t = 0^+$ is $i(0^+) = 0$ A.

38. Correct answer: **(C)**
This is a problem in the use of the rotational analog to Newton's Second Law of Motion, $\Sigma \vec{\tau} = I\vec{\alpha}$. Here there is a single torque

$$\tau = Fd = mgd = 4 \text{ kg}(10 \text{ m/sec}^2)(4 \text{ m}) = 160 \text{ Nm}$$

$$I = \frac{1}{2} MR^2 = \frac{1}{2}(40 \text{ kg})(4 \text{ m})^2 = 320 \text{ kg} \cdot \text{m}^2$$

So

$$\alpha = \frac{\tau}{I} = 0.5 \text{ rad/sec}^2$$

39. Correct answer: **(E)**
A pipe that supports a longitudinal sound wave will have a (displacement) node at a closed end and an antinode at an open end. This pipe has a node at one end and an antinode at the other. An infinite number of other nodes and antinodes might fall between these, but these correspond to smaller wavelength oscillations. The distance between an adjacent node and antinode is $\lambda/4$, so $\lambda/4 = 2$ m, or $\lambda = 8$ m.

40. Correct answer: **(B)**
In a liquid, the viscosity is related to the cohesion between the liquid particles. Therefore, because the cohesion decreased, the viscosity will decrease as the temperature increases.

41. Correct answer: **(B)**
We start from the formula $A_xB_yC_z$ and the relationships

$$x \cdot m_a = 2(y \cdot m_b)$$

and

$$x \cdot m_a = \frac{3}{2}(z \cdot m_c)$$

where x, y, and z must be integers. These give us $x = 3y$ and $x = 3z$, which are satisfied for $x = 3$, $y = 1$, $z = 1$, and integer multiples of these values. The smallest possible molecular mass is, therefore, $3(2) + 1(3) + 1(4) = 13$.

42. Correct answer: **(C)**
A real beam shows a linear stress–strain relationship for moderate loads and deviations from this for very large loads that induce substantial permanent deformations. Recall that stress refers to the cause (the load) and strain refers to the response (the deformation). Answer (B) is incorrect because it suggests that for increasing load there is a smaller deformation.

43. Correct answer: **(D)**
This problem may be solved graphically or by use of the thin lens equation

$$\frac{1}{d_o} + \frac{1}{d_i} = \frac{1}{f}$$

where d_o is the object distance, d_i the image distance, and f the focal length (positive for a converging lens, negative for a diverging lens). $d_o = 1.5$ m, $f = +0.5$ m, and we may solve for $d_i = +0.75$ m $= +3/2f$. Since this is positive, the image is on the "correct" side of the lens—the right side.

Graphically, we can draw any two of the principle rays; for example, we can draw the undeviated ray through the center of the lens and the ray from infinity that passes through the focal point:

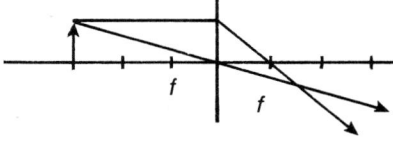

44. Correct answer: **(C)**

In a simple classical analysis, the radiation scattered by an object has the same frequency as the incident radiation. In the quantum mechanical problem, if a single photon transfers energy and momentum to a massive particle, it must lose energy itself. Since $E = h\nu$ for a photon, a loss in energy means a decrease in frequency, therefore an increase in wavelength. To compute the change in wavelength, apply conservation of energy and conservation of momentum.

45. Correct answer: **(C)**

The value of the sum of money P compounded annually at an interest rate i at the end of n years is

$$P = C(1 + i)^n$$

where C = initial sum of money.

For the given problem,

$$C = 20,000,$$
$$P = 2 \times 20,000 = 40,000$$
$$i = 0.11$$
$$n = \text{unknown years}$$

Therefore,

$$40,000 = 20,000(1.11)^n$$
$$2 = (1.11)^n$$
$$\ln 2 = \ln(1.11)^n$$
$$= n \ln(1.11)$$

and

$$n = \frac{\ln(2)}{\ln(1.11)} = \frac{0.6931}{0.1044} = 6.64$$

$n = 6.64$ years but since the money is compounded annually, it will take 7 years. At the end of 7 years, the 20,000 will have a value of

$$P = 20{,}000(1.11)^7 = \$41{,}523$$

**End of Solutions for Part A, Test 1.
Go to Solutions for Part B, Test 1.**

PART B

46. Correct answer: **(C)**
    Using the following fact:

    $$\int (u)^6 \frac{du}{dx} dx = \frac{(u)^7}{7} + c$$

    we write

    $$\int (\sin 2x)^6(\cos 2x) dx = \frac{1}{2} \int (\sin 2x)^6 (2\cos 2x) dx$$

    Therefore,

    $$\frac{1}{2} \frac{(\sin 2x)^7}{7} + c = \frac{1}{14} \sin^7 2x + c$$

47. Correct answer: **(A)**
    Write the constant multipliers separately. Thus,

    $$y = 2^2 x^2 + 3\sqrt{2}\sqrt{x} - \frac{1}{2}\frac{1}{x}$$

    $$= 4x^2 + 3\sqrt{2} x^{1/2} - \frac{1}{2} x^{-1}$$

    Then,

    $$\frac{dy}{dx} = 4(2x) + 3\sqrt{2} \frac{1}{2} x^{-1/2} - \frac{1}{2}(-1)x^{-2}$$

    Simplifying to obtain the answer gives

    $$\frac{dy}{dx} = 8x + \frac{3}{2}\sqrt{2} \frac{1}{\sqrt{x}} + \frac{1}{2}\frac{1}{x^2}$$

72

48. Correct answer: **(A)**

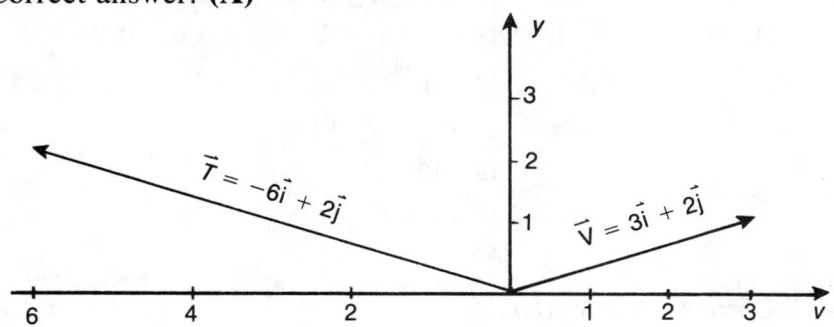

[Figure: Vectors $\vec{T} = -6\vec{i} + 2\vec{j}$ and $\vec{V} = 3\vec{i} + 2\vec{j}$ plotted on xy-axes]

Find the directed length of the vector $\vec{v}$ projected onto vector $\vec{T}$'s direction

$$\vec{v} \cdot \frac{\vec{T}}{|\vec{T}|} = (3i + 2j) \cdot \frac{(-6\hat{i} + 2\hat{j})}{\sqrt{(-6\vec{i} + 2\vec{j}) \cdot (-6\vec{i} + 2\vec{j})}}$$

$$= \frac{-18 + 4}{\sqrt{36 + 4}} = \frac{-14}{\sqrt{40}}$$

49. Correct answer: **(A)**

The rule for integrating a polynomial is

$$\int x^n \, dx = \frac{x^{n+1}}{n} \quad \text{when } n \neq -1$$

Therefore,

$$\int_0^1 (2x + 4x^3 - 3) \, dx = x^2 + x^4 - 3x \Big]_0^1$$

$$= 1 + 1 - 3 - 0 - 0 + 0 = -1$$

50. Correct answer: **(B)**

The product of a matrix $A = [a_{iq}]$ and a matrix $B = [b_{qj}]$ is (in the order) $AB = C$:

$$c_{ij} = a_{i1}b_{1j} + a_{i2}b_{2j} + \cdots + a_{iq}b_{qj} = \sum_{k=1}^{q} a_{ik}b_{kj}$$

$$(i = 1, 2, \cdots, m; j = 1, 2, \cdots, n)$$

where subscript i indicates the row and subscript j indicates the column.

$$\begin{vmatrix} 4 & 2 \\ 1 & 0 \end{vmatrix} \begin{vmatrix} 1 & 3 & 0 \\ 2 & 0 & 1 \end{vmatrix} = \begin{vmatrix} 4\times 1 + 2\times 2 & 4\times 3 + 2\times 0 & 4\times 0 + 2\times 1 \\ 1\times 1 + 0\times 2 & 1\times 3 + 0\times 0 & 1\times 0 + 0\times 1 \end{vmatrix}$$

$$= \begin{vmatrix} 4+4 & 12 & 2 \\ 1 & 3 & 0 \end{vmatrix}$$

$$= \begin{vmatrix} 8 & 12 & 2 \\ 1 & 3 & 0 \end{vmatrix}$$

51. Correct answer: **(B)**

    An identity matrix (I) is defined by a square matrix where all diagonal terms are 1 and all off diagonal terms are 0. Therefore,

    $$I = \begin{vmatrix} 1 & 0 \\ 0 & 1 \end{vmatrix} \text{ is an identity matrix}$$

52. Correct answer: **(B)**

    The most straightforward way to solve this problem is to multiply the expressions together. As a short cut, notice that there must be a constant term of 1 and a positive x^7 term. Only answer (B) satisfies this requirement. Alternately, the sum of the series

    $$1 + r + r^2 + \cdots + r^n$$

    is

    $$\frac{[1 - r^{n+1}]}{(1 - r)}$$

    Here, r is $-x$ and n is 6, so the sum is

    $$\frac{[1 - (-x)^7]}{(1 + x)}$$

    Multiplying this by $1 + x$ we obtain

    $$1 - (-x)^7 = 1 + x^7$$

53. Correct answer: **(C)**

    The product of the 3 by 2 matrix A with the 2 by 3 matrix B is the 3 by 3 matrix C,

    $$\begin{vmatrix} 13 & 11 & -1 \\ 11 & 17 & -7 \\ 5 & 5 & -1 \end{vmatrix}$$

The transpose of the matrix is obtained by the substitution $C_{ij} \to C_{ji}$. The resulting matrix is

$$\begin{vmatrix} 13 & 11 & 5 \\ 11 & 17 & 5 \\ -1 & -7 & -1 \end{vmatrix}$$

54. **Correct answer: (A)**
The operations of Gaussian elimination include multiplying a row by a scalar and adding or subtracting one row to or from another. Choice (B) can be obtained from matrix A by adding the first row to the second. Choice (E) can be obtained by multiplying the first row of A by 2. Only choice (A) cannot be obtained by these manipulations.

55. **Correct answer: (E)**
Cramer's rule states that the solution to the linear system

$$A_{11}X_1 + A_{12}X_2 + \cdots + A_{1n}X_n = B_1$$
$$\vdots \qquad \vdots \qquad \qquad \vdots \qquad \vdots$$
$$A_{n1}X_1 + A_{n2}X_2 + \cdots + A_{nn}X_n = B_n$$

is

$$X_1 = \frac{\begin{vmatrix} B_1 & A_{12} & \cdots & A_{1n} \\ B_2 & A_{22} & \cdots & A_{2n} \\ B_n & A_{n2} & \cdots & A_{nn} \end{vmatrix}}{\begin{vmatrix} A_{11} & \cdots & A_{1n} \\ A_{n1} & \cdots & A_{nn} \end{vmatrix}}$$

and similarly for the other X_i, where the ith column of the numerator matrix is replaced by the list of constants B_i. In this example, the denominator matrix is

$$\begin{vmatrix} 2 & 4 \\ 3 & -3 \end{vmatrix}$$

and the determinant is $2(-3) - 4(3) = -18$.

56. **Correct answer: (D)**
Of the 10 numbers in this set, 2 are even and 8 are odd. If the sum of the samples is even, either both numbers are even or both are odd.

There are $\boxed{2\,1}$ = 2 ways of picking a pair of even numbers (allowing for permutations) and $\boxed{8\,7}$ = 56 ways of picking a pair of odd numbers. In this sample space with 56 + 2 = 58 members there are $\boxed{1\,7}$ × 2 = 14 ways to pick a pair of numbers that includes 11. Thus the probability is

$$\frac{14}{58} \approx 0.24$$

57. Correct answer: **(D)**
A few modifications to the solution of Problem 56 are necessary here. There are $\boxed{2\,2}$ = 4 ways of picking a pair of even numbers and $\boxed{8\,8}$ = 64 ways of picking a pair of odd numbers for a sample space of 68 elements. There are $\boxed{1\,8}$ × 2 − 1 = 15 ways of picking an 11 because we do not want to double count the indistinguishable permutation of (11, 11). There is a probability of 15/68 ≈ 0.22 of picking an 11.

58. Correct answer: **(A)**
This pair of first-order differential equations is equivalent to the single second-order equation $\ddot{x} = -k\dot{x} - px$. This may be recognized from physics as Newton's Second Law of Motion applied to a damped oscillator. Solutions are damped exponentials of the form

$$\sin(\omega t)e^{-\mu t} \text{ or } e^{(a+ib)t}$$

Another approach is to try each of the answers in turn to see if it can solve the system. For

$$x = e^{(a+ib)t}, \dot{x} = (a + ib)e^{(a+ib)t} = (a + ib)x,$$
$$\ddot{x} = (a + ib)^2 e^{(a+ib)t} = (a + ib)^2 x$$

Plugging into the second-order equation, canceling the exponential, and grouping real and imaginary terms, we obtain two linear equations for a and b in terms of k and p. A solution of the complex exponential form can be found.

59. Correct answer: **(B)**
A Taylor series of $f(x)$ about $x = 0$ is

$$f(x) = f(0) + f'(0)\frac{x}{1!} + f''(0)\frac{x^2}{2!} + \cdots$$

Here,

$$f(0) = e^{-0^2} = 1, f'(0) = -2xe^{-x^2}\big|_0 = 0$$
$$f''(0) = 4x^2 e^{-x^2} - 2e^{-x^2}\big|_0 = -2$$

The Taylor series to second order is

$$e^{-x^2} \simeq 1 - x^2$$

(Alternately, e^u for small u is approximately $1 + u$, and we may substitute $-x^2$ for u.) The integral becomes

$$\int_0^{0.1} (1 - x^2) \, dx = x - \frac{x^3}{3}\bigg|_0^{0.1} = (0.1) - \frac{(0.1)^3}{3} = 0.0997$$

The Taylor series is likely to be a very good approximation to the function given the small upper limit of integration.

60. Correct answer: **(D)**
The indefinite integral of

$$\left(x - \frac{1}{x^2}\right)$$

is

$$\left(\frac{x^2}{2} + \frac{1}{x}\right)$$

using the rule

$$\int x^n \, dx = \frac{x^{n+1}}{n+1} + C$$

Substituting for x at the limits we have

$$\left(\frac{x^2}{2} + \frac{1}{x}\right)\bigg|_{1/y} - \left(\frac{x^2}{2} + \frac{1}{x}\right)\bigg|_y = \left(\frac{1}{2y^2} + y\right) - \left(\frac{y^2}{2} + \frac{1}{y}\right)$$

$$= \frac{1}{2y^2} + y - \frac{y^2}{2} - \frac{1}{y}$$

61. Correct answer: **(B)**
A useful trigonometric identity is

$$\sin^2(x) = \frac{1}{2} - \frac{1}{2}\cos(2x)$$

This is easily derived by subtracting

$$\cos(2x) = \cos^2(x) - \sin^2(x)$$

from

$$1 = \sin^2(x) + \cos^2(x)$$

(A similar half-angle formula for $\cos^2(x)$ results from adding these equations.)

$$\int_0^{2\pi} \sin^2(x)\, dx = \int_0^{2\pi} \left(\frac{1}{2} - \frac{1}{2}\cos(2x)\right) dx$$

$$= \left.\frac{x}{2}\right|_0^{2\pi} - \frac{1}{2}\int_0^{2\pi} \cos(2x)\, dx$$

The latter integral is 0 because it is the integral over two complete periods of the cosine function. The integral is therefore π.

62. **Correct answer: (B)**
This is a good candidate for a trigonometric substitution, in particular $1 + \tan^2\theta = \sec^2\theta$ [which is obtained from $\sin^2(\theta) + \cos^2(\theta) = 1$ after division by $\cos^2(\theta)$]. We make the substitution

$$x = \frac{1}{\sqrt{2}} \tan(\theta)$$

$$dx = \frac{1}{\sqrt{2}} \sec^2(\theta)\, d\theta$$

transforming the integral to

$$\frac{1}{\sqrt{2}} \int \frac{1}{1 + \tan^2\theta} \sec^2\theta\, d\theta = \frac{1}{\sqrt{2}} \int d\theta$$

63. **Correct answer: (B)**
The integral of the tangent function is not likely to be memorized, and there are not many tricks (integration by parts, trigonometric substitution, etc.) that can be used here. If we simply write $\tan(x)$ as

$$\frac{\sin(x)}{\cos(x)}$$

the integrand is seen to be of the form $-dy/y$ for $y = \cos(x)$. The integral is therefore $-\ln y = -\ln[\cos(x)]$. Working backward, each of the answers (A) to (D) can be differentiated to compare to $\tan(x)$.

64. Correct answer: **(B)**
The volume created here may be thought of as many thin disks stacked along the y axis. Each disk has area πr^2 and volume $\pi r^2\, dy$, where r is the distance from the y axis to the curve, or $r = x = y^{1/3}$. The total volume is then

$$\int_0^8 \pi r^2\, dy = \int_0^8 \pi y^{2/3}\, dy$$

End of Solutions for Part B, Test 1.
Go to Solutions for Part C, Test 1.

PART C

65. **Correct answer: (B)**
 If $x = g(t)$, then $dx = g'(t)\,dt$. The indefinite integral is transformed into $\int f[g(t)]g'(t)\,dt$. Finally, the integration limits $x = a$ and $x = b$ are transformed via $g(\alpha) = a$ and $g(\beta) = b$ into $\alpha = g^{-1}(a)$ and $\beta = g^{-1}(b)$.

66. **Correct answer: (D)**
 Integration by parts is usually written in the form

 $$\int u\,dv = uv - \int v\,du$$

 We first try $u = g'\,dx$, $dv = f'$ and obtain

 $$\int f'g'\,dx = g'f - \int fg''\,dx$$

 This is not one of the choices offered. If we try $u = f'$, $du = g'\,dx$ we obtain

 $$\int f'g'\,dx = f'g - \int f''g\,dx$$

 The trial-and-error approach (for the indefinite integral) may be tried here as well for simple functions f and g; for example, $f = x$ and $g = x^2$.

67. **Correct answer: (E)**
 The method of partial fractions is often used to simplify integrals where there is a product of polynomials in the denominator.

 $$\frac{1}{(ax+b)(cx^2+dx+e)} = \frac{\alpha}{ax+b} + \frac{\beta x + \delta}{cx^2+dx+e}$$

 can be solved for α, β, and δ by expanding

 $$\frac{1}{(ax+b)(cx^2+dx+e)} = \frac{\alpha(cx^2+dx+e) + (\beta x + \delta)(ax+b)}{(ax+b)(cx^2+dx+e)}$$

 $$= \frac{(\alpha c + \beta a)x^2 + [\alpha d + \delta a + b\beta]x + (\alpha e + \delta b)}{(ax+b)(cx^2+dx+e)}$$

 which provides the linear equations

 $$(\alpha c + \beta a) = 0,\quad (\alpha d + \delta a + \beta b) = 0,\quad (\alpha e + \delta b) = 1$$

We do not need to solve these here; we merely note that a solution exists. The integrals that remain are in (or can be manipulated into) the form

$$\int \frac{dx}{x}$$

for x a polynomial, which integrates to $\ln(x)$.

68. **Correct answer: (D)**
    Given a coordinate transformation

    $$x' = f(x, y), \; y' = g(x, y)$$

    the determinant of the Jacobi matrix

    $$\begin{vmatrix} \frac{\partial f}{\partial x} & \frac{\partial f}{\partial y} \\ \frac{\partial g}{\partial x} & \frac{\partial g}{\partial y} \end{vmatrix}_{(x_0, y_0)} = \left(\frac{\partial f}{\partial x} \frac{\partial g}{\partial y} - \frac{\partial f}{\partial y} \frac{\partial g}{\partial x} \right) \bigg|_{(x_0, y_0)}$$

    determines the local distortion of area elements. As an example, consider the transformation of polar coordinates to Cartesian coordinates. The defining transformation is $x = r\cos(\theta)$, $y = r\sin(\theta)$. We compute

    $$\frac{\partial f}{\partial r} = \cos\theta, \quad \frac{\partial f}{\partial \theta} = -r\sin\theta, \quad \frac{\partial g}{\partial r} = \sin\theta, \quad \frac{\partial g}{\partial \theta} = r\cos\theta$$

    $$\begin{vmatrix} \cos(\theta) & -r\sin(\theta) \\ \sin(\theta) & r\cos\theta \end{vmatrix} = r\cos^2(\theta) + r\sin^2\theta = r$$

    Hence, $dxdy = r(drd\theta)$.

69. **Correct answer: (C)**
    One approach is to compute

    $$g\left(\frac{1}{x}\right)$$

    and then take its first derivative:

    $$g\left(\frac{1}{x}\right) = \left(\frac{1}{x}\right)^2 + \frac{1}{(1/x)} = \frac{1}{x^2} + x$$

Hence,
$$g' = \frac{-2}{x^3} + 1$$

Alternately,
$$\frac{d}{dx}\{g[h(x)]\} = \frac{dg(h)}{dh}\frac{dh(x)}{dx}$$

Here $g = x^2 + 1/x$ and $h = 1/x$ so
$$\frac{dg(h)}{dh} = 2h - \frac{1}{h^2}$$

and
$$\frac{dg[h(x)]}{dh} = 2\left(\frac{1}{x}\right) - \frac{1}{(1/x)^2} = \frac{2}{x} - x^2, \quad \frac{dh(x)}{dx} = \frac{-1}{x^2}$$

Putting these together we have
$$\frac{d}{dx}\{g[h(x)]\} = \frac{-2}{x^3} + 1$$
$$= 1 - \frac{2}{x^3}$$

70. **Correct answer: (E)**
$g'(x)$ always negative implies $g(x)$ is everywhere decreasing, as is $f(x)$. The ratio of two such functions is not necessarily monotone since the two functions may decrease at different and varying rates. The ratio is not necessarily always positive or always negative since either decreasing function may be positive or negative.

71. **Correct answer: (A)**
While f is a periodic function, $F(x)$ is the total area under the curve f from 0 to x is therefore strictly increasing (since f is always positive).

72. **Correct answer: (E)**
The second derivative of a function is 0 only at an inflection point, a point where the function changes its concavity. Starting with concave up behavior, there are 6 concavity changes, as shown on the following page.

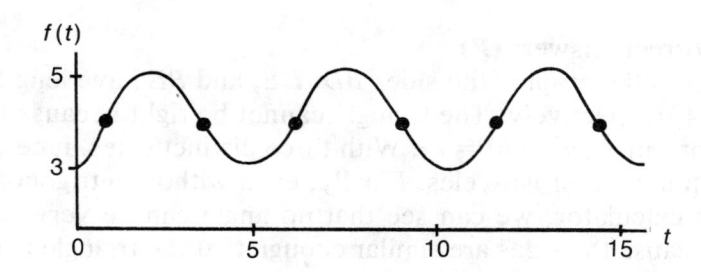

f(t) graph showing an oscillating curve between y=3 and y=5, with t from 0 to 15.

73. **Correct answer: (E)**
    $F(5)$ is the area under the curve from $t = 0$ to $t = 5$. The function f in this region looks somewhat like a triangle sitting on a rectangle. The triangle has base 5 and height 2, whereby its area is about $bh/2$, or 5, and the rectangle is 3 by 5 = 15, for a total area of 20.

74. **Correct answer: (D)**
    The derivative of the integral of a function is the function itself, so we can simply look for the minimum of the function f. The minimum of the function is 3.

75. **Correct answer: (C)**
    As we did in Problem 74, we can replace the words "derivative of F" with "the function f." Thus, we simply need the average of the function shown in the graph. The average appears to be 4 since the function is symmetric about $y = 4$.

76. **Correct answer: (B)**
    Object X starts at vertex A so if we look at the graph of distance from vertex B at $t = 0$ we find a separation of 5 units.

77. **Correct answer: (C)**
    As in Problem 76, we simply look at the graph of distance from vertex E and locate the plateau where X is sitting at B. This occurs at $t = 2.5$ sec. The distance is about 6.4 units.

78. **Correct answer: (A)**
    Since we have the graphs of distances to vertices B and E, we can find which of the remaining vertices is most nearly equidistant to these by reading off the lengths of AB, AE and CB, DE and DB, DE and FB, FE. AB and AE are both about 5, the closest pair of values.

79. Correct answer: **(E)**
From the graphs, the sides *BD*, *DE*, and *BE* have lengths 6, 5, and 6.4, respectively. The triangle cannot be right because the sides do not satisfy $a^2 + b^2 = c^2$. With three distinct sides, it certainly is not equilateral or isosceles. Finally, even without a trigonometric table or calculator, we can see that no angle can be very close to 180° because the sides are similar enough that the triangle is not very far from being equilateral (all 60° angles).

80. Correct answer: **(E)**
Each of the paths is composed of two segments, one end of which is either *B* or *E*. The total length for each path is therefore the sum of two distances read off of the graphs. For example, path *BEC* is composed of segment *BE* of length 6.4 and segment *EC* of length about 6.3.

81. Correct answer: **(B)**
The integral from 0 to x of a p.d.f. is the cumulative density function (c.d.f.) at x. The c.d.f. represents the total probability up to x and must therefore asymptote to 1. The p.d.f. rises rapidly and then falls to 0, so its integral must rise rapidly and then level off at 1.

82. Correct answer: **(C)**
A change in concavity means a transition between concave up (second derivative positive) and concave down behavior (second derivative negative). This must occur when the second derivative is 0. The p.d.f. is concave down from $x = 0$ to $x \approx 1.3$ and concave up from $x = 1.3$ to $x = 3$.

83. Correct answer: **(E)**

$$\frac{d}{dx}[\tan(x)\sin(x)] = \frac{d}{dx}\left(\frac{\sin^2(x)}{\cos(x)}\right)$$

so we have the option of using the quotient rule or the product rule. By the product rule,

$$\frac{d}{dx}[\tan(x)\sin(x)] = \tan'(x)\sin(x) + \tan(x)\sin'(x)$$

$$= \sec^2(x)\sin(x) + \tan(x)\cos(x)$$

which cannot be brought into any of the given forms by trigonometric identities. By the quotient rule

$$\frac{d}{dx}\left(\frac{\sin^2(x)}{\cos(x)}\right) = \frac{\cos^2(x)2\sin(x) + \sin^3(x)}{\cos^2(x)} = 2\sin(x) + \tan^2(x)\sin(x)$$

which also cannot be transformed into any of the functions given.

84. Correct answer: **(D)**
    The derivative of c^x is $c^x \ln(c)$. (As a check of this, recall that e^x is its own derivative.) If we replace x by $f(x)$, then by the chain rule the above result is multiplied by $f'(x)$. Here $f(x) = x^2$ and $c = 2$, so we have $2^{x^2}(\ln 2)(2x)$.

85. Correct answer: **(C)**
    A quantity increasing by $x\%$ a total of y times has a new value of $(1 + x)^y$ times its former value for a percent increase of

    $$[(1 + x)^y - 1] \times 100\%$$

    In the case of 5% annual interest compounded monthly, each increase is by 0.05/12.

End of Solutions for Part C, Test 1.
Go to Solutions for Part D, Test 1.

PART D

86. **Correct answer: (B)**
Each of the incident particles has a total energy that reflects its rest energy mc^2 and its momentum. Conservation of mass and energy tells us that this energy becomes the rest energy *only* Mc^2 of the composite mass, since this object has no motion in the center of mass frame. Clearly $M > 2m$.

87. **Correct answer: (A)**
Equivalent capacitor is

$$\frac{1}{C_{eq}} = \frac{1}{C_1} + \frac{1}{C_2} = \frac{1}{3} + \frac{1}{6} = \frac{1}{2}$$

or

$$C_{eq} = 2\ \mu F$$

Charge stored in C_{eq} is

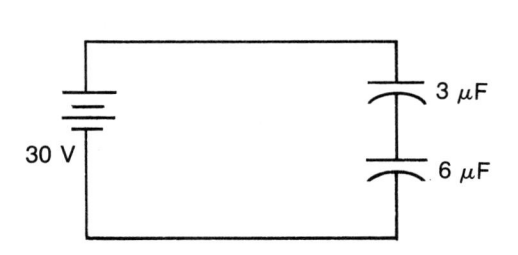

$$Q = C_{eq} \cdot V = 2 \times 30 = 60\ \mu C$$

This charge is stored in each capacitor; that is

$$Q_{6\ \mu F} = Q_{3\ \mu F} = 60\ \mu C$$

88. **Correct answer: (B)**
Equilibrium occurs when the change in free energy G is 0. For the reaction

$$aA + bB = cC + dD,$$

$$\Delta G = \Delta G° + RT \ln \left(\frac{a_C^c \times a_D^d}{a_A^a \times a_B^b}\right)$$

where $\Delta G°$ is the free energy change in the standard state, and the a's are the activities of each chemical species. Thus,

$$\Delta G° = -RT \ln \left(\frac{a_C^c \times a_D^d}{a_A^a \times a_B^b}\right)$$

and K is defined by $\Delta G° = -RT \ln K$. The activity of a gas is given by its partial pressure times a factor that is close to 1 at low pressures. Answer (B) then obtains.

It is also possible to replace the partial pressures by the concentrations of each species, but

$$K = \frac{C_{HCl}^2}{C_{H_2} C_{Cl_2}}$$

was not offered as a choice.

89. **Correct answer: (D)**
The current in a series RLC circuit is maximum when the circuit is at resonance. The resonance occurs when

$$X_L = X_C$$

Hence,

$$X_C = 30 \; \Omega$$

90. **Correct answer: (B)**

$$\text{Force} = \text{pressure} \cdot \text{area}$$
$$\text{Pressure} = \gamma \cdot h$$

where γ = specific weight of liquid.
Therefore,

$$P_{AA'} = \gamma \cdot 0 = 0$$
$$P_{BB'} = \gamma \cdot 2h = 2\gamma h$$
$$P_{CC'} = \gamma \cdot h = \gamma h$$
$$P_{DD'} = \gamma \cdot 3h = 3\gamma h$$

and

$$F_{AA'} = 0 \cdot a = 0$$
$$F_{BB'} = 2h \cdot 3a = 6\gamma h a$$
$$F_{CC'} = h \cdot 2a = 2\gamma h a$$
$$F_{DD'} = 3h \cdot a = 3\gamma h a$$

Therefore, the largest force is at BB'

91. **Correct answer: (D)**
This is a problem of electromagnetic induction. The voltage across the coil is given by

$$V = N \frac{\Delta \phi}{\Delta t} = 240 \times \frac{50 \text{ Wb/min}}{60 \text{ sec/min}} = 200 \text{ V}$$

88 / Graduate Record Examination in Engineering

The current through the coil is

$$i = \frac{V}{R} = \frac{200 \text{ V}}{40 \text{ }\Omega} = 5 \text{ A}$$

92. Correct answer: **(C)**
    The total vertical force on the inclined surface is acting upward. The magnitude of this force is the same as the imaginary weight of liquid above the inclined surface at a height h.

    $$F_v = \gamma \times \text{volume} = \gamma \times \frac{(h \times h \times W)}{2} = \frac{\gamma h^2 W}{2}$$

93. Correct answer: **(A)**
    The basic assumptions of an ideal gas are as follows:

    (i) any finite volume of a gas consists of a very large number of molecules

    (ii) as in (A)

    (iii) the molecules are in continuous motion

    (iv) molecules exert no forces on one another except when they collide

    (v) collision of molecules with one another and with the walls are perfectly elastic

    (vi) in the absence of external forces, the molecules are distributed uniformly throughout the container

    (vii) All directions of molecular velocities are assumed equally probable

94. Correct answer: **(A)**
    Light reflected from the top of the plate undergoes a π phase shift because it is reflected off of an optically denser medium. The transmitted light has no phase shift at any of the interfaces it encounters. The difference in optical path length at the top of the plate is therefore π (for the reflected ray) + $2\pi(2t/\lambda')$ (for the other ray). The wavelength of light within the glass plate is $\lambda' = \lambda/1.5$. Complete destructive interference requires that the optical path length differ-

ence is an odd multiple of π. The smallest thickness solution (other than ϕ) requires

$$\pi + 2\pi \frac{3t}{\lambda} = 3\pi \text{ or } t = \frac{\lambda}{3}$$

95. **Correct answer: (A)**
   According to the kinetic theory of gases, the absolute temperature T of an ideal monatomic gas is related to the mean square speed of the molecules as follows:

   $$\frac{3}{2} kT = \frac{1}{2} m\overline{v^2}$$

   where

   k = Boltzmann constant
   m = mass of a molecule
   $\overline{v^2}$ = mean square speed

96. **Correct answer: (C)**
   In (C) we see the three phases of water at moderate pressures: liquid, gaseous, and ordinary ice. At constant pressure, we can make a transition from ice to water to vapor, or from ice directly to vapor, and so on. We also see a point of coexistence of the three phases. In (E) the phase diagram for water at high pressures is shown with its different solid phases.

97. **Correct answer: (C)**
   A polarizing sheet (ideally) transmits all of the electric field parallel to its polarizing axis and none of the electric field at right angles to this axis. Unpolarized light incident on a plane polarizer therefore becomes perfectly plane polarized and loses one-half of its intensity. A second polarizer parallel to the first would have no effect on the transmitted light; one at right angles to the first would cut out all of the light. Clearly, it is the projection of the first polarizer's axis on the second that determines the amplitude of the transmitted light. This is a factor of $\cos(\theta)$. The intensity of the light is proportional to the amplitude squared, or $\cos^2 \theta$. Hence

   $$I = \frac{I_0}{2} \cos^2 \theta$$

   and if

   $$I = \frac{1}{6} I_0$$

then

$$\theta = \cos^{-1}\left(\frac{1}{\sqrt{3}}\right)$$

98. Correct answer: **(B)**
The equation of continuity of a fluid is a statement of mass conservation. If a fluid is incompressible, then the equation reduces to the statement that the volume of fluid passing by any cross section per unit time is a constant. The volume per unit time is the velocity times the area, $A|\vec{v}|$. If the fluid is compressible, the density varies, and mass conservation requires an integral over the density variations.

99. Correct answer: **(D)**
The motion of a cylinder rolling down a plane can be determined from energy considerations or by use of Newton's Second Law of Motion in translational and rotational forms. As a shortcut, however, we should recall that an object in free-fall is governed by the equations

$$v = at, \ a = g, \ s = at^2/2$$

Putting these together, we find

$$v_{\text{final}} = \sqrt{2gh}$$

that is, the final velocity depends only on the vertical distance traveled. When we recall that the inclined plane historically has been used to reduce the effective value of g, the result obtains.

100. Correct answer: **(D)**
Each 5730 years that pass decrease the carbon-14 concentration by a factor of 2. Six percent of the original concentration means about 1 part in 16, or $(1/2)^4$, so four half-lives represents the oldest sample that can be accurately dated. Answer (D) is very close to four half-lives.

101. Correct answer: **(C)**
The de Broglie wavelength of a massive particle is h/p, where h is Planck's constant and p is the particle's momentum. If the mass is doubled and the velocity halved, there is no change in the nonrelativistic momentum mv and therefore no change in the wavelength.

102. Correct answer: **(B)**
Since the height of water in the tank is constant and we are neglecting all losses,

$$\text{Flow in} = \text{flow out}$$
$$100 \text{ ft}^3/\text{sec} = vA$$

where A = area of the hole in the wall.
Therefore,

$$v = \frac{100 \text{ ft}^3/\text{sec}}{2.5 \text{ ft}^2} = 40 \text{ ft/sec}$$

103. Correct answer: **(D)**
If an atom has only closed shells, then electron spins cancel and there is no total angular momentum. A closed $n = 1$ shell means a $1s^2$ configuration, and requires exactly two electrons. A closed $n = 2$ shell means a $1s^2 2s^2 2p^6$ configuration, and requires 10 electrons.

104. Correct answer: **(A)**
From the figure below, the result obtains if you mentally erase each of the chlorine atoms.

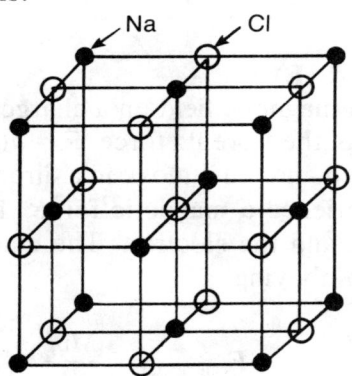

105. Correct answer: **(A)**
The internal energy, U, of a closed pVT system may be written as

$$U = U(T, V)$$

or

$$dU = \left(\frac{\partial U}{\partial T}\right)_V dT + \left(\frac{\partial U}{\partial V}\right)_T dV$$
$$= C_v dT + \left(\frac{\partial U}{\partial V}\right)_T dV$$

since

$$C_v = \left(\frac{\partial U}{\partial T}\right)_V$$

For an ideal gas, for all temperatures and pressures,

$$pV = RT \text{ and } \left(\frac{\partial U}{\partial V}\right)_T = 0$$

Thus, for an ideal gas, for all temperatures and pressures,

$$dU = C_v dT$$

106. Correct answer: **(D)**

$$\text{Annual depreciation cost} = \frac{\text{first cost} - \text{salvage value}}{\text{life in years}}$$

Therefore,

$$\text{annual depreciation cost} = \frac{\$20,000 - \$2,000}{10} = \$1,800$$

107. Correct answer: **(E)**
The effect of a magnetic field on a charged particle comes from the magnetic part of the Lorentz force, $\vec{F} = q(\vec{v} \times \vec{B})$. Clearly, a particle must be charged and must move in a direction other than the B field if it is to experience a magnetic force. The only charged particles here are alpha and the electron. The charged particle will move in a circular path obeying

$$F_{\text{centripetal}} = \frac{Mv^2}{R} = qvB$$

so

$$R = \frac{Mv^2}{qB}$$

A particle is most affected if its radius of curvature is small (the trajectory is sharply bent). The radius is proportional to M/q so the very light electron should be most deviated from its original motion.

108. Correct answer: **(C)**
The binding energy per nucleon must be greater in the final state in each case. For fusion, we go from very low to somewhat higher masses; that is, we travel the curve in the positive x direction from the low end, so the function must start monotone up. For fission, we

start at high mass values and travel to much lower masses, so the function must grow as we travel in the negative x direction from the high end.

109. Correct answer: **(A)**
According to the kinetic theory of gases, the pressure p is given by

$$p = \frac{1}{3} mn\overline{v^2}$$

where

m = mass of the molecule
n = the number density
$\overline{v^2}$ = the mean square velocity

Therefore, the pressure p varies linearly as the number density n. The absolute temperature T is given by

$$\frac{3}{2} kT = \frac{1}{2} m\overline{v^2}$$

The absolute temperature is independent of the number density n.

110. Correct answer: **(D)**

$$\text{Pressure} = \text{specific weight} \times \text{depth}$$

The pressure drops as you go vertically upward in a liquid, and the pressure increases as you go vertically down. Starting at P_A,

$$P_A - \bar{\gamma}L - \gamma L + \bar{\gamma}L = P_B$$

Therefore,

$$P_A - P_B = \gamma L$$

111. Correct answer: **(B)**
The specific heats at constant pressure and volume are properties of the substance and are always positive.

112. Correct answer: **(E)**
Due to the pull of gravity the pressure in a static container is

$$\text{Pressure} = \gamma \times \text{height of water}$$

If the effect of the pull of gravity is zero as in the case of a free-falling object, the pressure is zero throughout the body and, therefore, zero at the bottom of the bucket.

113. Correct answer: **(C)**
The maximum frictional force is (coefficient of friction) · (force) = 0.15 * 120 N = 18 N. The application of Newton's Second Law of Motion gives us $F - 18 = ma = 12(3)$. Solving for F, we find a force of 54 N will produce the desired acceleration.

114. Correct answer: **(B)**
The position of a mass on a spring is

$$x(t) = A \cos(\omega t)$$

From successive differentiation, the velocity and acceleration expressions are sine and cosine functions, respectively, and are therefore 90° out of phase. If you remember that the acceleration is zero when the velocity is a maximum and vice versa, the same result is obtained.

115. Correct answer: **(B)**
The amount of solar energy impinging on a surface placed normal to the sun's irradiation decreases with increasing distances from the sun. In fact, it varies inversely with the square of the distance from the sun; that is, if Q is the amount of solar energy impinging on a surface that is at a distance of R away from the sun,

$$Q \propto \frac{1}{R^2}$$

Thus, assuming $Q_{s.v.}$ as the solar energy impinging on the space vehicle (s.v.), and Q_{earth} as the solar energy impinging on the earth's surface,

$$\frac{Q_{s.v.}}{Q_{earth}} = \frac{1}{\left(\frac{2}{3}\right)^2}$$

$$Q_{s.v.} = \frac{9}{4} \times 1.4 \times 10^3 \, \frac{J}{\text{sec} \cdot m^2}$$

$$= 3.2 \times 10^3 \, \frac{J}{\text{sec} \cdot m^2}$$

Sample Test 1 / 95

116. Correct answer: **(C)**
The downward forces on the submerged object are its weight, 600 N, and the external force, 200 N, for a total of 800 N. If it is in equilibrium, the net upward force must also be 800 N. The buoyancy force is equal to the weight of the displaced fluid, that is, the volume of the displaced fluid times its density times g.

$$800 \text{ N} = V \left(\frac{1000 \text{ kg}}{\text{m}^3}\right)\left(\frac{10 \text{ m}}{\text{sec}^2}\right)$$

Solving for V, which is the same as the volume of the object, we have 0.08 m³.

117. Correct answer: **(B)**
This problem can be solved without any calculation. The force in the southwest direction has a component along the negative y axis that cancels (at least approximately by eye) the entire first force. What is left of the $\overline{SW}$ force is a component to the west that can be canceled by a force to the east. Only answer (B) appears to work.
To solve analytically, break each force into x and y components, sum the components along each axis, and combine the components to find the net force. The third force is equal in magnitude and opposite in direction to the net force.

118. Correct answer: **(B)**
Due to constant uniform horizontal acceleration in an open container, the liquid will adjust itself to move as a solid. From the equation of motion in the horizontal direction

$$\sum F_x = M a_x$$
$$F_x = P + \Delta P - P = \Delta P A$$

where A is the cross-sectional area of the arbitrary horizontal segment cross hatched and P = pressure and

$$M = \frac{\text{weight}}{g} = \frac{\gamma L A}{g}$$

where γ = specific weight. Therefore,

$$\Delta P A = \frac{\lambda L A}{g} a_x$$

and

$$\frac{a_x}{g} = \frac{\Delta P}{\lambda L} \qquad (1)$$

Sum forces in vertical direction

$$\sum F_y = Ma_y$$

and since

$$a_y = 0$$
$$\sum F_y = 0$$

that is hydrostatic pressure or

$$\Delta P = \gamma \Delta h \qquad (2)$$

Substituting (2) into (1)

$$\frac{a_x}{g} = \frac{\Delta h}{L} = \tan \alpha$$

and the hydrostatic pressure at point A increases.

119. **Correct answer: (E)**
Thevenin's equivalent to the generator is given below

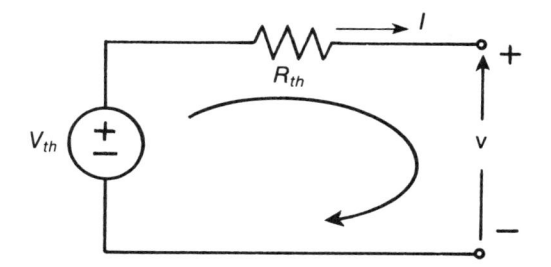

and the voltage regulation is defined by

$$\text{Regulation} = \frac{V_{\text{no-load}} - V_{\text{full-load}}}{V_{\text{full-load}}} \times 100$$

$$5 = \frac{V_{nl} - 5}{5} \times 100$$

$$V_{nl} = 5.25 \text{ V} = V_{th}$$

Writing *KVL* in the loop,

$$V_{th} = V + IR_{th}$$

At no load,

$$I = 0, V = V_{th} = 5.25 \text{ V}$$

At full load,

$$I = 4 \text{ A}, \quad V = 5 \text{ V}$$
$$5.25 = 5 + 4R_{th}$$

or

$$R_{th} = 0.0625 \text{ }\Omega$$

120. Correct answer: **(A)**
For every action there is an equal and opposite reaction. In somewhat more detail, each force from object A to object B has a "mirror" force from object B to object A, equal in magnitude, opposite in direction—along the line of the first force. The apparent paradox of the two forces canceling out is resolved by remembering that the forces act on two distinct objects, not just on the cart. Recall Newton's Second Law: The sum of all external forces acting on a (single) mass is equal to the product of that mass and its resulting acceleration.

121. Correct answer: **(C)**
Consider a bathroom-type scale. It is being pushed down by your weight and being pushed up by the floor, yet it reads only your weight. Or consider what happens in the system shown if one of the weights is removed and the loose string is tied to the end of the table. This corresponds to the usual way that a spring scale is used, yet the tied string experiences the same force (the force required to maintain static equilibrium) as if the weight were there. In a string the tension is the same at all points, and a spring scale measures the tension at a *point*.

122. Correct answer: **(B)**
We can ignore the numerical factors and reduce the expression to

$$\frac{\eta \mu R^4}{h^3}$$

Further, since R and h both have the units of length this reduces to $\eta \mu R$ or

$$\left(\frac{\text{g}}{\text{cm} \cdot \text{sec}} \times \frac{\text{cm}}{\text{sec}} \times \text{cm}\right) = \frac{\text{g} \cdot \text{cm}}{\text{sec}^2}$$

These are the same units as

$$\text{force} = \text{mass} \times \text{acceleration} = (g) \left(\frac{\text{cm}}{\text{sec}^2}\right)$$

123. **Correct answer: (D)**

The work done when an ideal gas expands isothermally and reversibly in a piston-and-cylinder assembly is

$$W = RT \ln \left(\frac{p_i}{p_f}\right)$$

$$= 8314 \frac{J}{\text{kg-mol}°K} \cdot 300°K \ln \left(\frac{10}{1}\right)$$

$$= 5{,}743{,}000 \frac{J}{\text{kg-mol}}$$

The work done by a pressure acting through a volume, V, as in the case of a fluid pressure exerted on a piston, is

$$W = \int p \, dV$$

Since the process is mechanically reversible and the fluid is an ideal gas,

$$P = \frac{RT}{V}$$

Thus,

$$W = \int_{V_i}^{V_f} RT \frac{dV}{V} = RT \ln \left(\frac{V_f}{V_i}\right)$$

Since

$$P_i V_i = P_f V_f$$

$$W = RT \ln \left(\frac{P_i}{P_f}\right)$$

124. **Correct answer: (D)**

$\sum$ Horizontal reaction $= \rho A v (v_{out} - v_{in})$

For this problem,

$v = v_{in}$

$v_{out} = 0$ in x direction

$\sum F_x = \left(2 \frac{\text{slugs}}{\text{ft}^3}\right)\left(\frac{1 \text{ in.}^2}{144 \text{ in.}^2/\text{ft}^3}\right)(12)\left(0 - 12 \frac{\text{ft}^2}{\text{sec}}\right)$

$\sum F_x = -2 \text{ lb}$

The minus sign implies the reaction for F_x is in the negative x direction. Therefore,

$$\text{horizontal force} = 2 \text{ lb}.$$

125. Correct answer: **(C)**
From Bernoulli's equation,

$$Z_1 + \frac{v_1^2}{2g} + \frac{P_1}{\gamma} = Z_2 + \frac{v_2^2}{2g} + \frac{P_2}{\gamma}$$

For a horizontal pipe,

$$Z_1 = Z_2$$

and given

$$v_2 = 2v_1, \quad \gamma = 2g$$

$$\frac{v_1^2}{2g} + \frac{P_1}{2g} = \frac{(2v_1)^2}{2g} + \frac{P_2}{2g}$$

$$P_2 - P_1 = v_1^2 - 4v_1^2 = -3v_1^2$$

or the drop in pressure is $3v_1^2$.

126. Correct answer: **(D)**
From Bernoulli's equation

$$Z_1 + \frac{v_1^2}{2g} + \frac{P_1}{\gamma} = Z_2 + \frac{v_2^2}{2g} + \frac{P_2}{\gamma}$$

Take the datum at the center of the discharge, therefore,

$$Z_2 = 0$$

For a reservoir the surface is open to the atmosphere and remains static. Therefore,

$$Z_1 = H, \quad P_1 = 0, \quad \text{and } v_1 = 0$$

At the discharge $P_2 = 0$ since the flow is open to the atmosphere. Therefore,

$$H = \frac{v_2^2}{2g}$$

$$v = \sqrt{2gH}$$

127. Correct answer: **(C)**

For steady flow, the velocity is not a function of time t; therefore, for three-dimensional steady flow

$$v = f(x, y, z)$$

For one-dimensional flow, the velocity (and other flow parameters—pressure, density, etc.) is a function of the coordinate that is parallel to the stream lines, for this example, the x coordinate. The coordinates perpendicular to the stream lines are held constant, y and z. If the velocity does not vary in the x direction, the flow is uniform and the velocity is constant. In this example, the flow may accelerate across a reducer, therefore,

$$v = f(x)$$

End of Solutions for Part D, Test 1.
Go to Solutions for Part E, Test 1.

PART E

128. **Correct answer: (D)**
The normal distribution (or bell curve) is the function

$$\frac{1}{\sigma\sqrt{2\pi}} \exp\left\{\frac{-(x-\mu)^2}{2\sigma^2}\right\}$$

where μ is the mean and σ is the variance (the square of the standard deviation). Clearly this curve is sharply peaked around the mean (relative to σ) so the mean ± 2 standard deviations will contain the majority of the data.

129. **Correct answer: (B)**
The Taylor series expansion for $f(x)$ about $x = 0$ is

$$f(0) + f'(0)\frac{x}{1!} + f''(0)\frac{x^2}{2!} + \cdots$$

Here

$$f(0) = \tan(0) = 0, f'(0) = \sec^2(0) = \frac{1}{\cos^2(0)} = 1$$

so the first term of the series is x. If you recall the Taylor series for $\sin(x)$ (x + cubic and higher corrections) and $\cos(x)$ (1 + quadratic and higher corrections) the ratio is seen to be x to at least first order.

130. **Correct answer: (D)**
We can immediately take the limit $n \to \infty$ and consider

$$\sum_{0}^{\infty} \frac{1}{3^k}$$

This is a geometric series with first term 1 and multiplier 1/3. The infinite sum is in general the first term divided by 1 minus the multiplier, or

$$\frac{1}{1-(1/3)} = \frac{3}{2}$$

This can quickly be derived by writing the sum as

$$S = 1 + \frac{1}{3} + \frac{1}{9} + \cdots$$

recognizing $(1/3)S$ as $S - 1$ and solving for S.

131. Correct answer: **(C)**
This expression is composed of the two geometric series

$$1 + \frac{1}{2} + \frac{1}{4} + \frac{1}{8} + \cdots$$

and

$$\frac{1}{3} + \frac{1}{9} + \frac{1}{27} + \cdots$$

Each is of the form

$$\sum_{n=0}^{\infty} ar^n$$

whose sum is

$$\frac{a}{1-r}$$

In the first series,

$$\frac{1}{1 - (1/2)} = 2$$

In the second series

$$\frac{1/3}{1 - (1/3)} = \frac{1}{2}$$

The expression sums to 2 1/2.

132. Correct answer: **(B)**
Any linear expression in three dimensions is a two-dimensional plane. The intersection of two planes is null if they are parallel, a plane if they are the same plane, or else a line. The planes here are clearly not the same, and we can verify that they are not parallel by showing that there is at least one point of intersection. From the first

equation

$$y = 7 - x$$

and from the second equation

$$z - 4x + 5(7 - x) = 0$$

These equations have a solution, proving that there is an intersection; in fact, these equations define the line in which they meet.

133. Correct answer: **(B)**
By substituting $x = 2$ into $x^2 + y^2 = z$ we obtain

$$y^2 = z - 4,$$

which is the equation of a parabola parallel to the y, z plane.

134. Correct answer: **(B)**
First, we must write the linear relationship between z and x from the given information. Writing $z = ax + b$, we have the two equations

$$x = ax + b, \quad -24 = -24a + b$$

and

$$-x = ax + b, \quad -8 = 8a + b$$

Solving these we find

$$a = \frac{1}{2}, \quad b = -12$$

Now, by simple substitution

$$y = 3\left(\frac{z - b}{a}\right) + 60 = 6z + 132 \text{ or } z = \frac{y}{6} - 22$$

135. Correct answer: **(A)**

$$\frac{e^x}{x^x} = \left(\frac{e}{x}\right)^x$$

is a small number raised to a large power (for large x) and therefore goes to zero from above. $\sin(x)$ oscillates about zero but is finite for all x. Their product

$$\sin(x) \cdot \left(\frac{e}{x}\right)^x$$

goes to zero in an oscillatory (rather than a monotone) manner.

136. Correct answer: **(C)**

One approach here is to choose convenient values for x and/or y and evaluate each of the expressions. For example, if $y = 1$, then

$$\ln(x^y) = \ln(x^1) = \ln x$$

This can be compared to

$$\ln y \ln x = \ln(1) \ln x = 0$$
$$(\ln x)^y = (\ln x)^1 = \ln x$$
$$y \ln x = \ln x$$

and

$$x \ln y - x \ln 1 = 0$$

Unfortunately, answers (B) and (C) are both consistent with this particular choice of y. Further experimentation would point to answer (C). A more analytical approach is to write x^y as

$$(e^{\ln x})^y = e^{y \ln x}$$

and then take the natural log of both sides. Finally, if b was an integer

$$a^b = \overbrace{a \cdot a \cdot a \cdots a}^{b}$$

we could use the property

$$\ln \prod_{i=1}^{n} = \sum_{i=1}^{n} \ln$$

137. Correct answer: **(B)**

Euler's identity reads

$$e^{i\theta} = \cos(\theta) + i \sin(\theta)$$

Plugging in $\theta = 2\pi$ we have

$$\cos(2\pi) + i\sin(2\pi) = \cos(0) + i\sin(0) = 1 + i0 = 1$$

More simply, $e^{i2\pi n}$ is the same for all n (periodic with period 2π), including $n = 0$, and $e^0 = 1$.

138. **Correct answer: (B)**
In general, the behavior of such a sequence may be quite complicated. However, the only answers offered are fixed points, defined by the condition

$$X_{n+1} = X_n$$

in the limit n goes to infinity. Setting

$$X_{n+1} = X_n$$

the recursive prescription becomes $X_n^2 = 5$

therefore $\qquad X_n = \sqrt{5}$

139. **Correct answer: (E)**
The speed of an object in three-dimensional space is

$$s = \sqrt{\frac{dx^2}{dt} + \frac{dy^2}{dt} + \frac{dz^2}{dt}}$$

Here

$$\frac{dx}{dt} = 2t$$

$$\frac{dy}{dt} = 1$$

$$\frac{dz}{dt} = -4t$$

so

$$s = \sqrt{(2t)^2 + (1)^2 + (-4t)^2} = \sqrt{1 + 20t^2}$$

Speed is not to be confused with the *vector* quantity velocity, choice (A).

140. Correct answer: **(C)**

The standard deviation is the square root of the variance and is therefore 3. The highest score is 8 points above the mean, which corresponds to 8/3 standard deviations.

**End of Solutions for Test 1.
Go to Test 2.**

Sample Test 2
Answer Sheet

1 Ⓐ Ⓑ Ⓒ Ⓓ Ⓔ	29 Ⓐ Ⓑ Ⓒ Ⓓ Ⓔ	57 Ⓐ Ⓑ Ⓒ Ⓓ Ⓔ	85 Ⓐ Ⓑ Ⓒ Ⓓ Ⓔ	113 Ⓐ Ⓑ Ⓒ Ⓓ Ⓔ
2 Ⓐ Ⓑ Ⓒ Ⓓ Ⓔ	30 Ⓐ Ⓑ Ⓒ Ⓓ Ⓔ	58 Ⓐ Ⓑ Ⓒ Ⓓ Ⓔ	86 Ⓐ Ⓑ Ⓒ Ⓓ Ⓔ	114 Ⓐ Ⓑ Ⓒ Ⓓ Ⓔ
3 Ⓐ Ⓑ Ⓒ Ⓓ Ⓔ	31 Ⓐ Ⓑ Ⓒ Ⓓ Ⓔ	59 Ⓐ Ⓑ Ⓒ Ⓓ Ⓔ	87 Ⓐ Ⓑ Ⓒ Ⓓ Ⓔ	115 Ⓐ Ⓑ Ⓒ Ⓓ Ⓔ
4 Ⓐ Ⓑ Ⓒ Ⓓ Ⓔ	32 Ⓐ Ⓑ Ⓒ Ⓓ Ⓔ	60 Ⓐ Ⓑ Ⓒ Ⓓ Ⓔ	88 Ⓐ Ⓑ Ⓒ Ⓓ Ⓔ	116 Ⓐ Ⓑ Ⓒ Ⓓ Ⓔ
5 Ⓐ Ⓑ Ⓒ Ⓓ Ⓔ	33 Ⓐ Ⓑ Ⓒ Ⓓ Ⓔ	61 Ⓐ Ⓑ Ⓒ Ⓓ Ⓔ	89 Ⓐ Ⓑ Ⓒ Ⓓ Ⓔ	117 Ⓐ Ⓑ Ⓒ Ⓓ Ⓔ
6 Ⓐ Ⓑ Ⓒ Ⓓ Ⓔ	34 Ⓐ Ⓑ Ⓒ Ⓓ Ⓔ	62 Ⓐ Ⓑ Ⓒ Ⓓ Ⓔ	90 Ⓐ Ⓑ Ⓒ Ⓓ Ⓔ	118 Ⓐ Ⓑ Ⓒ Ⓓ Ⓔ
7 Ⓐ Ⓑ Ⓒ Ⓓ Ⓔ	35 Ⓐ Ⓑ Ⓒ Ⓓ Ⓔ	63 Ⓐ Ⓑ Ⓒ Ⓓ Ⓔ	91 Ⓐ Ⓑ Ⓒ Ⓓ Ⓔ	119 Ⓐ Ⓑ Ⓒ Ⓓ Ⓔ
8 Ⓐ Ⓑ Ⓒ Ⓓ Ⓔ	36 Ⓐ Ⓑ Ⓒ Ⓓ Ⓔ	64 Ⓐ Ⓑ Ⓒ Ⓓ Ⓔ	92 Ⓐ Ⓑ Ⓒ Ⓓ Ⓔ	120 Ⓐ Ⓑ Ⓒ Ⓓ Ⓔ
9 Ⓐ Ⓑ Ⓒ Ⓓ Ⓔ	37 Ⓐ Ⓑ Ⓒ Ⓓ Ⓔ	65 Ⓐ Ⓑ Ⓒ Ⓓ Ⓔ	93 Ⓐ Ⓑ Ⓒ Ⓓ Ⓔ	121 Ⓐ Ⓑ Ⓒ Ⓓ Ⓔ
10 Ⓐ Ⓑ Ⓒ Ⓓ Ⓔ	38 Ⓐ Ⓑ Ⓒ Ⓓ Ⓔ	66 Ⓐ Ⓑ Ⓒ Ⓓ Ⓔ	94 Ⓐ Ⓑ Ⓒ Ⓓ Ⓔ	122 Ⓐ Ⓑ Ⓒ Ⓓ Ⓔ
11 Ⓐ Ⓑ Ⓒ Ⓓ Ⓔ	39 Ⓐ Ⓑ Ⓒ Ⓓ Ⓔ	67 Ⓐ Ⓑ Ⓒ Ⓓ Ⓔ	95 Ⓐ Ⓑ Ⓒ Ⓓ Ⓔ	123 Ⓐ Ⓑ Ⓒ Ⓓ Ⓔ
12 Ⓐ Ⓑ Ⓒ Ⓓ Ⓔ	40 Ⓐ Ⓑ Ⓒ Ⓓ Ⓔ	68 Ⓐ Ⓑ Ⓒ Ⓓ Ⓔ	96 Ⓐ Ⓑ Ⓒ Ⓓ Ⓔ	124 Ⓐ Ⓑ Ⓒ Ⓓ Ⓔ
13 Ⓐ Ⓑ Ⓒ Ⓓ Ⓔ	41 Ⓐ Ⓑ Ⓒ Ⓓ Ⓔ	69 Ⓐ Ⓑ Ⓒ Ⓓ Ⓔ	97 Ⓐ Ⓑ Ⓒ Ⓓ Ⓔ	125 Ⓐ Ⓑ Ⓒ Ⓓ Ⓔ
14 Ⓐ Ⓑ Ⓒ Ⓓ Ⓔ	42 Ⓐ Ⓑ Ⓒ Ⓓ Ⓔ	70 Ⓐ Ⓑ Ⓒ Ⓓ Ⓔ	98 Ⓐ Ⓑ Ⓒ Ⓓ Ⓔ	126 Ⓐ Ⓑ Ⓒ Ⓓ Ⓔ
15 Ⓐ Ⓑ Ⓒ Ⓓ Ⓔ	43 Ⓐ Ⓑ Ⓒ Ⓓ Ⓔ	71 Ⓐ Ⓑ Ⓒ Ⓓ Ⓔ	99 Ⓐ Ⓑ Ⓒ Ⓓ Ⓔ	127 Ⓐ Ⓑ Ⓒ Ⓓ Ⓔ
16 Ⓐ Ⓑ Ⓒ Ⓓ Ⓔ	44 Ⓐ Ⓑ Ⓒ Ⓓ Ⓔ	72 Ⓐ Ⓑ Ⓒ Ⓓ Ⓔ	100 Ⓐ Ⓑ Ⓒ Ⓓ Ⓔ	128 Ⓐ Ⓑ Ⓒ Ⓓ Ⓔ
17 Ⓐ Ⓑ Ⓒ Ⓓ Ⓔ	45 Ⓐ Ⓑ Ⓒ Ⓓ Ⓔ	73 Ⓐ Ⓑ Ⓒ Ⓓ Ⓔ	101 Ⓐ Ⓑ Ⓒ Ⓓ Ⓔ	129 Ⓐ Ⓑ Ⓒ Ⓓ Ⓔ
18 Ⓐ Ⓑ Ⓒ Ⓓ Ⓔ	46 Ⓐ Ⓑ Ⓒ Ⓓ Ⓔ	74 Ⓐ Ⓑ Ⓒ Ⓓ Ⓔ	102 Ⓐ Ⓑ Ⓒ Ⓓ Ⓔ	130 Ⓐ Ⓑ Ⓒ Ⓓ Ⓔ
19 Ⓐ Ⓑ Ⓒ Ⓓ Ⓔ	47 Ⓐ Ⓑ Ⓒ Ⓓ Ⓔ	75 Ⓐ Ⓑ Ⓒ Ⓓ Ⓔ	103 Ⓐ Ⓑ Ⓒ Ⓓ Ⓔ	131 Ⓐ Ⓑ Ⓒ Ⓓ Ⓔ
20 Ⓐ Ⓑ Ⓒ Ⓓ Ⓔ	48 Ⓐ Ⓑ Ⓒ Ⓓ Ⓔ	76 Ⓐ Ⓑ Ⓒ Ⓓ Ⓔ	104 Ⓐ Ⓑ Ⓒ Ⓓ Ⓔ	132 Ⓐ Ⓑ Ⓒ Ⓓ Ⓔ
21 Ⓐ Ⓑ Ⓒ Ⓓ Ⓔ	49 Ⓐ Ⓑ Ⓒ Ⓓ Ⓔ	77 Ⓐ Ⓑ Ⓒ Ⓓ Ⓔ	105 Ⓐ Ⓑ Ⓒ Ⓓ Ⓔ	133 Ⓐ Ⓑ Ⓒ Ⓓ Ⓔ
22 Ⓐ Ⓑ Ⓒ Ⓓ Ⓔ	50 Ⓐ Ⓑ Ⓒ Ⓓ Ⓔ	78 Ⓐ Ⓑ Ⓒ Ⓓ Ⓔ	106 Ⓐ Ⓑ Ⓒ Ⓓ Ⓔ	134 Ⓐ Ⓑ Ⓒ Ⓓ Ⓔ
23 Ⓐ Ⓑ Ⓒ Ⓓ Ⓔ	51 Ⓐ Ⓑ Ⓒ Ⓓ Ⓔ	79 Ⓐ Ⓑ Ⓒ Ⓓ Ⓔ	107 Ⓐ Ⓑ Ⓒ Ⓓ Ⓔ	135 Ⓐ Ⓑ Ⓒ Ⓓ Ⓔ
24 Ⓐ Ⓑ Ⓒ Ⓓ Ⓔ	52 Ⓐ Ⓑ Ⓒ Ⓓ Ⓔ	80 Ⓐ Ⓑ Ⓒ Ⓓ Ⓔ	108 Ⓐ Ⓑ Ⓒ Ⓓ Ⓔ	136 Ⓐ Ⓑ Ⓒ Ⓓ Ⓔ
25 Ⓐ Ⓑ Ⓒ Ⓓ Ⓔ	53 Ⓐ Ⓑ Ⓒ Ⓓ Ⓔ	81 Ⓐ Ⓑ Ⓒ Ⓓ Ⓔ	109 Ⓐ Ⓑ Ⓒ Ⓓ Ⓔ	137 Ⓐ Ⓑ Ⓒ Ⓓ Ⓔ
26 Ⓐ Ⓑ Ⓒ Ⓓ Ⓔ	54 Ⓐ Ⓑ Ⓒ Ⓓ Ⓔ	82 Ⓐ Ⓑ Ⓒ Ⓓ Ⓔ	110 Ⓐ Ⓑ Ⓒ Ⓓ Ⓔ	138 Ⓐ Ⓑ Ⓒ Ⓓ Ⓔ
27 Ⓐ Ⓑ Ⓒ Ⓓ Ⓔ	55 Ⓐ Ⓑ Ⓒ Ⓓ Ⓔ	83 Ⓐ Ⓑ Ⓒ Ⓓ Ⓔ	111 Ⓐ Ⓑ Ⓒ Ⓓ Ⓔ	139 Ⓐ Ⓑ Ⓒ Ⓓ Ⓔ
28 Ⓐ Ⓑ Ⓒ Ⓓ Ⓔ	56 Ⓐ Ⓑ Ⓒ Ⓓ Ⓔ	84 Ⓐ Ⓑ Ⓒ Ⓓ Ⓔ	112 Ⓐ Ⓑ Ⓒ Ⓓ Ⓔ	140 Ⓐ Ⓑ Ⓒ Ⓓ Ⓔ

Sample Test 2

Time—170 Minutes
140 Questions

PART A

Directions: Each of the problems or incomplete statements below is followed by five suggested answers or completions. Select the one that is best in each case and then blacken the corresponding space on the answer sheet.

1. Two coils have an equivalent inductance of 20 H if the connection is in series and an equivalent inductance of 12 H if the connection is opposing. Determine the mutual inductance M.

   (A) 32 H
   (B) 8 H
   (C) 4 H
   (D) 2 H
   (E) 0 H

2. A sphere weighs 10 lb in air and is submerged in a container of liquid with a specific weight of 70 lb/ft^3. If the volume of the sphere is 0.1 ft^3, find the reading on the scale, W.

   (A) 17 lb
   (B) 10 lb
   (C) 7 lb
   (D) 3 lb
   (E) 0 lb

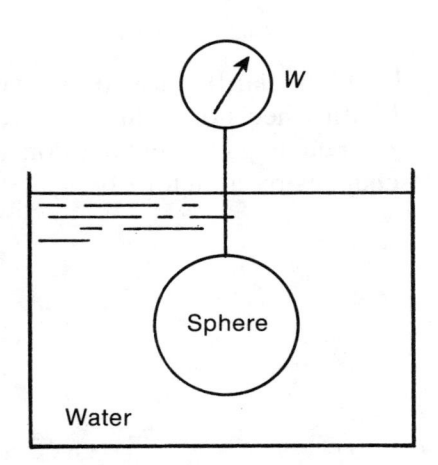

3. Given the following circuit with an ideal transformer

    If $I_1 = 0.2$ A and $I_2 = 20$ A, determine the turn ratio $\dfrac{n_1}{n_2}$

    (A) $\dfrac{200}{1}$

    (B) $\dfrac{100}{1}$

    (C) $\dfrac{10}{1}$

    (D) $\dfrac{1}{10}$

    (E) $\dfrac{1}{100}$

4. The temperature of an ideal gas is doubled. By what factor does the average (root-mean-square) speed of the component molecules increase?

    (A) 1
    (B) $\sqrt{2}$
    (C) 2
    (D) 4
    (E) none of the above

5. The angular momentum of a system is conserved if there

    (A) are no forces present
    (B) are no magnetic forces present
    (C) is no net force on the system
    (D) are no torques present
    (E) is no net torque acting on the system

6. A bicycle of (standard) rear wheel drive has its wheels of diameter d replaced with wheels of diameter $d' < d$ in front and $d'' > d$ in the back. If the pedaling rate remains constant, the linear velocity of the bicycle changes by a factor of

    (A) $\dfrac{d''}{d}$

    (B) $\dfrac{d''}{d'}$

    (C) $\dfrac{(d'' - d)}{(d' - d)}$

(D) $\dfrac{(d' - d)}{(d'' - d)}$

(E) $\dfrac{(d' + d'')}{(2d)}$

7. Monochromatic light strikes an air–glass interface from the air side at a 1° angle of incidence. The index of refraction of the glass is 1.5. The deviation of the transmitted light from the original direction is about

   (A) 0°
   (B) 0.33° toward the surface normal
   (C) 0.33° away from the surface normal
   (D) 0.5° toward the surface normal
   (E) 0.66° away from the surface normal

8. A system whose transfer function has been given in Laplace transform domain is stable if

   (A) all the poles are on the left half of the S plane
   (B) all the poles remain within the unit circle
   (C) all the zeros are on the left half of the S plane
   (D) the poles of the jw axis are of the multiplicity of two or more
   (E) the zeros on the jw axis are simple zeros

9. The frictionless spring–mass system shown below oscillates once every 2 sec. If the mass m is 4 kg, what is the spring constant k?

   (A) 2 N/m
   (B) 4 N/m
   (C) $\dfrac{\pi^2}{16}$ N/m
   (D) $4\pi^2$ N/m
   (E) π^2 N/m

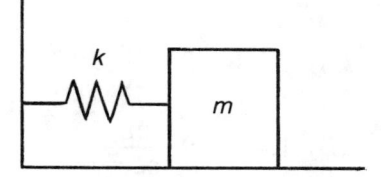

10. How much work is done by an ideal gas when it undergoes a reversible isothermal expansion from volume V_1 to volume V_2?

    (A) $p(V_2 - V_1)$
    (B) $nRT\left(\dfrac{V_2 - V_1}{V_1}\right)$

(C) $nRT \ln \dfrac{V_2}{V_1}$

(D) no work is done

(E) none of the above

11. For the circuit shown below, the voltage v_x is most nearly

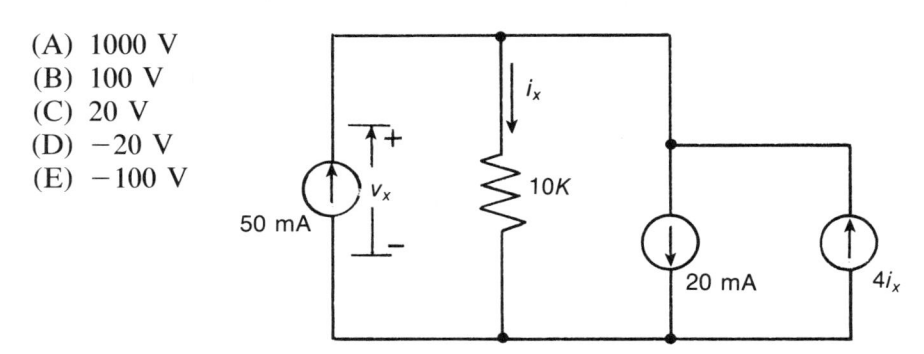

(A) 1000 V
(B) 100 V
(C) 20 V
(D) −20 V
(E) −100 V

12. A flashbulb generates a hemispherical wave front of light as measured by a person holding the bulb. A second person, at the same point in space but moving at 80% of the speed of light, measures

(A) the same hemisphere
(B) a hemisphere of smaller radius
(C) a hemisphere of larger radius
(D) an ellipsoidal shape
(E) none of the above

13. An electron of mass 9.11×10^{-31} kg and charge 1.6×10^{-19} C is accelerated through 1200 V from its initial speed of 10^7 m/sec. What is its approximate final velocity (assuming classical mechanics applies)?

(A) $1.2 \times 10^7 \dfrac{\text{m}}{\text{sec}}$

(B) $1.4 \times 10^7 \dfrac{\text{m}}{\text{sec}}$

(C) $1.8 \times 10^7 \dfrac{\text{m}}{\text{sec}}$

(D) $2.2 \times 10^7 \dfrac{\text{m}}{\text{sec}}$

(E) $2.9 \times 10^7 \dfrac{\text{m}}{\text{sec}}$

14. A laser produces intense, monochromatic, coherent light by which of the following processes?

    (A) stimulated absorption
    (B) stimulated emission
    (C) spontaneous emission
    (D) photoelectric effect
    (E) bremsstrahlung

15. The critical point of water is reached at a critical temperature of 705°F, a critical pressure of 3206.2 lbf/in.^{2}, and a critical volume of 0.0503 ft^3/lbm. Which one of the following statements is true?

    (A) there is a constant-temperature vaporization process
    (B) the saturated-liquid and saturated-vapor states are identical
    (C) there is no definite change in phase from liquid to vapor
    (D) there is no definite change in phase from solid to liquid
    (E) there is no definite change in phase from solid to vapor

16. There are several reasons why a lens may not produce a sharp image of an object point. Which one of the following is *not* a valid reason?

    (A) the incident light is not monochromatic
    (B) the image is being formed at the wrong point in space
    (C) the lens is diverging
    (D) the paraxial approximation is being violated
    (E) the object is as large as the lens that images it

17. Given the circuit below, at $t = 0$ both sources turned off as indicated. Determine the current i_L at $t = 0^+$.

    (A) 16 A
    (B) 4 A
    (C) −4 A
    (D) −16 A
    (E) −20 A

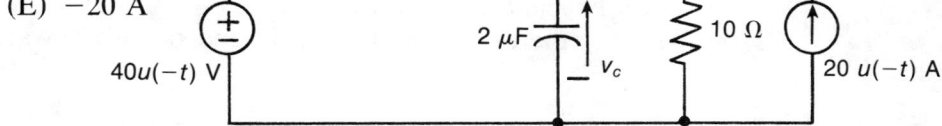

    40u(−t) V, 10 H, 2 μF, V_c, 10 Ω, 20 u(−t) A

18. A person in a cold and windy environment loses body heat through the processes of conduction, convection, and radiation. Which of these processes is (are) likely to be insignificant compared to the others?

    (A) conduction

(B) convection
(C) radiation
(D) convection and radiation
(E) conduction and radiation

19. The quantum state of an electron in a hydrogen atom is specified by the numbers n, l, m_l, and m_s. The number l is most closely related to

(A) the average radius of the electron's orbit
(B) the direction of the electron's intrinsic spin
(C) the eccentricity of the electron's orbit
(D) the orientation of the electron's orbit in three dimensions
(E) none of the above

20. In the circuit given below, the switch S is closed for a long time. The switch S is opened at $t = 0$. Determine the power supply voltage E_0 such that $V(t) = 1$ V at $t = 0.1$ sec.

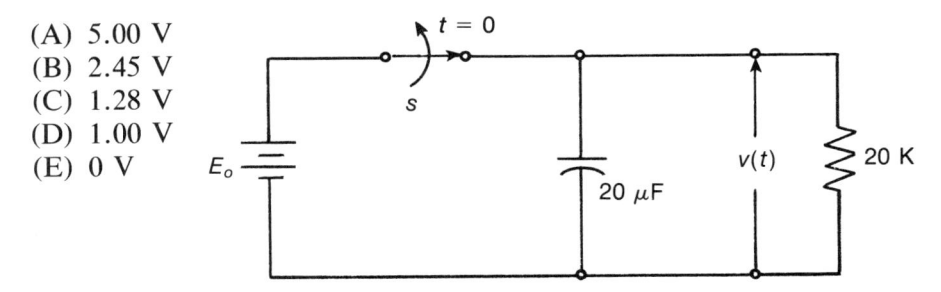

(A) 5.00 V
(B) 2.45 V
(C) 1.28 V
(D) 1.00 V
(E) 0 V

21. An infinitely long current-carrying wire points in the same direction as a constant electric field, and the current flows in the same direction as the electric field. A positive test charge placed at rest near the wire will

(A) remain at rest
(B) move at constant velocity
(C) have a constant acceleration along the wire
(D) exhibit a helical (screwlike) motion
(E) move along the direction of the wire and radially toward it

22. A liquid is irregularly stirred in a well-insulated container and thereby undergoes a rise in temperature. Regarding the liquid as the system, which of the following statements correctly describes the process?

(A) heat has been transferred
(B) work has been done

(C) there is no change in internal energy of the system
(D) it is an isentropic process
(E) none of the above

23. A pair of coupled coils are shown below.
Where $L_1 = 4$ H, $L_2 = 6$ H, $M = 1$ H, what is the value of a single inductor that will replace L_1 and L_2?

(A) 7 H
(B) 8 H
(C) 10 H
(D) 11 H
(E) 12 H

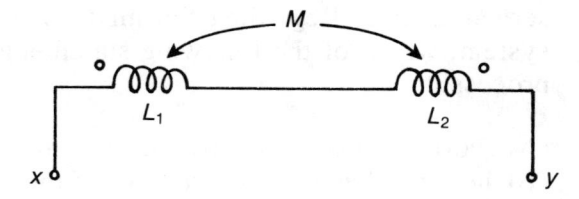

Problems 24 and 25 are based on the following information.

An object moves along a spiral in three-dimensional space at constant speed. The axis of the spiral is the z axis, the radius of the spiral is R, and $x = z = 0$ at $t = 0$; $x = R$, $z = 3R$ at $t = 2$ sec. The parametric equations describing the motion are of the form

$$x = R \cos(\omega t + \phi)$$
$$y = R \sin(\omega t + \phi)$$
$$z = kt$$
$$\text{where } k = \frac{3R}{2}$$

24. The value of ω is:

(A) 0

(B) $\frac{\pi}{2}$

(C) π

(D) 2π

(E) cannot be determined from given information

25. The magnitude of the object's velocity at any time t is given by

(A) k

(B) $k + R\omega$

(C) ωR

(D) $\sqrt{k^2 + (R\omega)^2}$

(E) $\dfrac{\omega R}{k}$

26. A combustion experiment is performed by burning a mixture of fuel and oxygen in a constant volume "bomb" surrounded by a water bath. During the experiment, the temperature of the water is observed to rise. Regarding the mixture of fuel and oxygen as the system, which of the following statements correctly describes the process?

    (A) work has been done by the system
    (B) heat has been transferred out of the system
    (C) there is no change in internal energy of the system
    (D) it is an isentropic process
    (E) it is an adiabatic process

27. A person swims straight across a stream with velocity $\vec{v}_L$ (as viewed by someone on land) despite a current of $\vec{v}_W$ downstream. To swim straight across, the swimmer had a velocity vector $\vec{v}_S$ (relative to the water) satisfying

    (A) $\vec{v}_S = \vec{v}_L - \vec{v}_W$
    (B) $\vec{v}_S = \vec{v}_L + \vec{v}_W$
    (C) $|\vec{v}_S| = |\vec{v}_L| + |\vec{v}_W|$
    (D) $|\vec{v}_S| < |\vec{v}_W|$
    (E) none of the above

28. An electric dipole consists of a positive and negative charge of equal magnitude separated by a fixed distance. Which of the following is *not* a property of the dipole's electric field?

    (A) field lines close on the dipole
    (B) field strength decreases as $\dfrac{1}{r^3}$ for large r
    (C) field strength decreases as $\dfrac{1}{r^2}$ for large r
    (D) the field is symmetric about the line joining the charges
    (E) the field is nearly radial very close to either charge

29. Fifty grams of steam at 100°C are cooled to liquid water at 80°C. How much heat is transferred in the process (heat capacity of water is 1 cal/(g °C), heat of vaporization is 540 cal/g)?

(A) 28,000 cal
(B) 11,500 cal
(C) 2,800 cal
(D) 1,000 cal
(E) none of the above

30. A two-slit interference pattern, ignoring the superimposed one-slit diffraction patterns, looks most like (ink indicates a *bright* region):

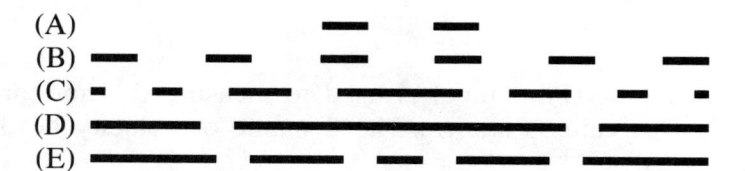

(A)
(B)
(C)
(D)
(E)

The truss shown below applies to Problems 31 and 32.

31. Find the force in member *AB*.

(A) 20.0 lb
(B) 14.1 lb
(C) 10.0 lb
(D) 7.10 lb
(E) 0 lb

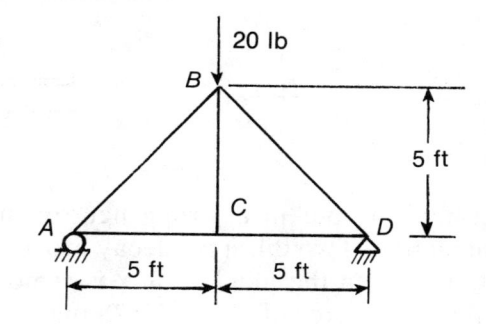

32. Find the force in member *AC*.

(A) 20.0 lb
(B) 14.1 lb
(C) 10.0 lb
(D) 7.1 lb
(E) 0 lb

33. For a closed system, the difference between the heat added to the system and the work done by the system is equal to the change in

(A) enthalpy
(B) entropy
(C) temperature
(D) internal energy
(E) heat capacity

34. As atomic numbers increase in the periodic table, atomic masses increase faster. Which one of the following makes the greatest contribution to this extra mass?

   (A) for each new proton there is a new electron mass
   (B) additional nuclear binding energy
   (C) additional atomic binding energy
   (D) additional molecular binding energy
   (E) extra neutrons are required to overcome the increasing proton–proton repulsion

35. An L-shaped tube (Pitot tube) is used to measure the horizontal velocity in a stream as shown. The dynamic or velocity head is (neglect losses)

   (A) H
   (B) $\Delta h + H$
   (C) Δh
   (D) 0
   (E) $H - \Delta h$

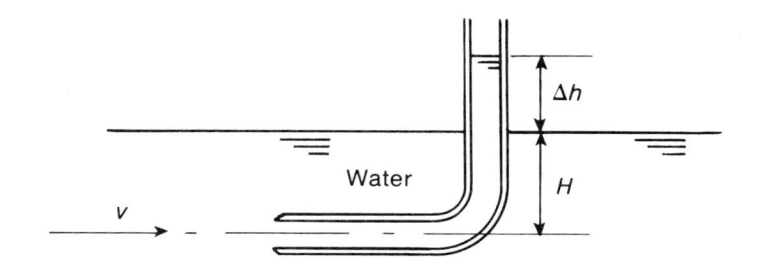

36. A sample is put in a strong neutron flux to make it temporarily radioactive. Two different decay processes are observed with equal activity when the incident flux is removed and with activities that differ by a factor of $e = 2.71828$ one hour later. What is the difference in the decay rates of the two radioactive species?

   (A) one hr^{-1}
   (B) 10^{10} hr
   (C) e sec
   (D) $\frac{1}{e}$ sec^{-1}
   (E) 1 sec^{-1}

37. The beam shown on the following page is free to pivot about point A. A 10-lb vertical force is applied at one end of the beam and a cube is hanging off the opposite end. The cube is submerged in water with a specific gravity of 62.4 lb/ft^3 and the density of the block is 112.4 lb/ft^3. Find the dimension of the side of the cube to keep the system as shown in equilibrium.

(A) 32 ft

(B) 8 ft

(C) 1 ft

(D) $\frac{1}{2}$ ft

(E) $\frac{1}{8}$ ft

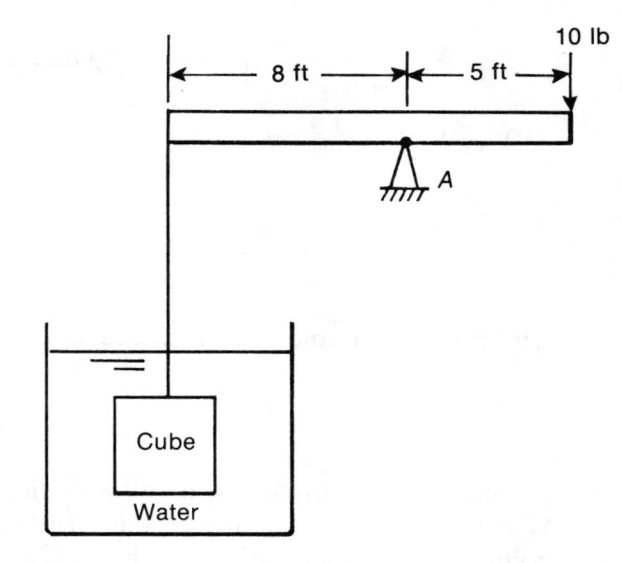

38. Which of the following statements about the self-inductance of a solenoid is true?

   (A) it depends on the current through it
   (B) it depends on the voltage across it
   (C) it depends on details of its construction, not on current or voltage
   (D) it depends on the proximity of a second solenoid
   (E) it does not change if an iron core is inserted

39. Natural gallium consists of two isotopes whose weights are 68.93 and 70.92. What is the percentage abundance of the heavier isotope if a sample has an atomic weight of 69.724 g?

   (A) 39.9%
   (B) 50.4%
   (C) 60.1%
   (D) 69.2%
   (E) 76.7%

40. With the switch closed, the circuit shown on the following page is in a steady state for $t < 0$. The switch opens at $t = 0$. Determine the voltage across the switch at $t = 0^+$, $v_S(0^+)$.

(A) $\frac{24}{11}$ V

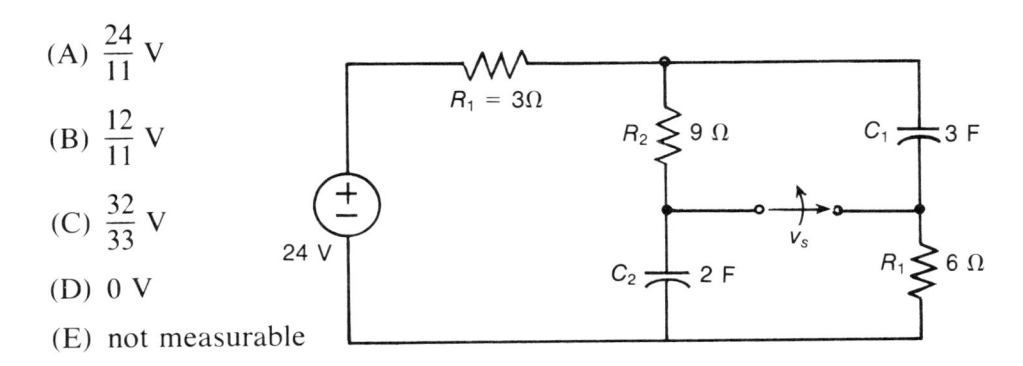

(B) $\frac{12}{11}$ V

(C) $\frac{32}{33}$ V

(D) 0 V

(E) not measurable

41. A sphere rotates uniformly about a vertical axis through its center. Which one of the following best describes how the linear speed of a point on the surface varies with latitude as one travels from equator to pole? (The latitude is zero at the equator.)

    (A) it decreases as first power of latitude
    (B) it decreases as cos(latitude)
    (C) it is constant
    (D) it increases as sin(latitude)
    (E) it decreases as cos²(latitude)

42. A rigid massless bar has a length of 4 m. At distances of 1, 2, 3, and 4 m from the left end, 1-, 2-, 3-, and 4-kg masses are placed, respectively. How far from the left end will a support need to be placed to balance the load?

    (A) 1.5 m
    (B) 2.4 m
    (C) 3 m
    (D) 3.2 m
    (E) 3.6 m

43. Transition elements differ from other elements in their greater tendency to

    (A) be chemically inert
    (B) fill internal subshells as the atomic number increases
    (C) act chemically identical to all other transition elements
    (D) have no net angular momentum
    (E) none of the above

44. When sodium and chlorine are brought together each attains a rare gas configuration. This bond is

   (A) covalent
   (B) ionic
   (C) a hydrogen bond
   (D) a form of hybridization
   (E) magnetic in origin

45. How much would $1,000 be worth at the end of 3 years if it were drawing interest at a rate of 10.0% a year, compounded twice a year?

   (A) $1,771
   (B) $1,340
   (C) $1,331
   (D) $1,300
   (E) $1,000

End of Part A.
Go to Part B.

PART B

Directions: For each problem, select the best of the choices offered. Computation and scratch work may be done in this examination book.

46. If

    $$G(x) = \frac{x}{x+2},$$

    find

    $$G\left(\frac{1}{x}\right)$$

    (A) $\frac{1}{x}$

    (B) $\frac{1}{2+x}$

    (C) $\frac{1}{1+2x}$

    (D) $2x + 1$

    (E) $1 + 2x$

47. Find the rank of the following matrix

    $$\begin{vmatrix} 1 & 2 & 1 \\ 3 & 3 & 0 \\ 1 & 1 & 0 \end{vmatrix}$$

    (A) 4
    (B) 3
    (C) 2
    (D) 1
    (E) 0

48. $\int_0^{\pi/12} \sin 3x \, dx =$

   (A) $\dfrac{\sqrt{2}}{3}$

   (B) $1 - \sqrt{2}$

   (C) $\dfrac{1 - \dfrac{1}{\sqrt{2}}}{3}$

   (D) $-\dfrac{1}{3}$

   (E) 0

49. $\dfrac{d}{dx} \ln(2x + 3)$

   (A) $\dfrac{1}{2x + 3}$

   (B) $\dfrac{2x + 3}{2}$

   (C) $\dfrac{2}{2x + 3}$

   (D) e^{2x+3}

   (E) $2 \ln(2x + 3)$

50. $\dfrac{d}{dx} (e^x \sin x) =$

   (A) $e^x \cos x$

   (B) $\dfrac{\cos x}{x}$

   (C) $e^x(\sin x + \cos x)$

   (D) e^x

   (E) $e^x \sin^2 x$

51. $\int_0^{1/x} g'(t)dt =$

   (A) $g(0)$

   (B) $g\left(\dfrac{1}{x}\right) - g(0)$

(C) $g(0) - g\left(\frac{1}{x}\right)$

(D) $\frac{1}{x}$

(E) 0

52. Suppose

$$f(x) = \begin{cases} x^2 \text{ for } -1 \leq x \leq 1 \\ 0 \text{ else} \end{cases}$$

and suppose

$$F(x) = c \int_{-1}^{x} f(t) dt$$

What value does the constant c have if F is a cumulative distribution function?

(A) π

(B) 1

(C) $\frac{3}{2}$

(D) $\frac{1}{2}$

(E) 0

53. The differential equation

$y''(x) = ky(x)$

(k is a constant) has which of the following kind(s) of nontrivial solutions?

(A) sinusoidal functions only
(B) exponential functions only
(C) hyperbolic functions only
(D) sinusoids and exponentials
(E) polynomials of order 2

54. Which of the following matrices does *not* have an inverse?

    (A) $\begin{pmatrix} 3 & 6 \\ 6 & 11 \end{pmatrix}$

    (B) $\begin{pmatrix} 4 & 9 \\ -4 & 9 \end{pmatrix}$

    (C) $\begin{pmatrix} -1 & 5 \\ -2 & 10 \end{pmatrix}$

    (D) $\begin{pmatrix} 1 & 0 \\ 0 & 1 \end{pmatrix}$

    (E) $\begin{pmatrix} 2 & 5 \\ -1 & 4 \end{pmatrix}$

55. The product of $(1 \ \ 2 \ \ 5)$ with $\begin{pmatrix} 2 \\ 4 \\ 7 \end{pmatrix}$ is

    (A) not defined
    (B) 45
    (C) $(2 \ \ 8 \ \ 35)$
    (D) $\begin{pmatrix} 2 \\ 8 \\ 35 \end{pmatrix}$
    (E) 0

56. In the figure below, the line segment *AB* bisects the angle *CAD*. What is the length of line segment *AE*?

    (A) $\dfrac{AB}{2}$

    (B) $\dfrac{AB \cdot AD}{AC}$

    (C) $\dfrac{DE \cdot AD}{BC}$

    (D) $\sqrt{\dfrac{AC \cdot AD}{DE}}$

    (E) $AB - \left(\dfrac{AD}{DE}\right)$

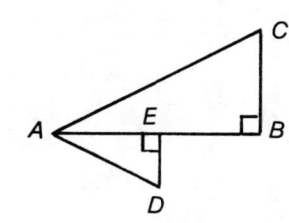

57. Which one of the following matrices is an orthogonal matrix (a matrix whose transpose equals its inverse)?

(A) $\begin{pmatrix} \frac{2}{5} & \frac{4}{5} \\ \frac{-1}{5} & 1 \end{pmatrix}$

(B) $\begin{pmatrix} 4 & -7 \\ 2 & 0 \end{pmatrix}$

(C) $\begin{pmatrix} \frac{1}{\sqrt{2}} & \frac{1}{\sqrt{2}} \\ -\frac{1}{\sqrt{2}} & \frac{1}{\sqrt{2}} \end{pmatrix}$

(D) $\begin{pmatrix} \frac{1}{6} & \frac{5}{6} \\ \frac{2}{3} & \frac{1}{3} \end{pmatrix}$

(E) $\begin{pmatrix} 1 & 0 \\ 1 & -1 \end{pmatrix}$

58. $|\vec{A} - \vec{B}|$ is

(A) $= |\vec{A}| + |\vec{B}|$
(B) $\geq |\vec{A}| + |\vec{B}|$
(C) $\leq |\vec{A}| + |\vec{B}|$
(D) $< |\vec{A}| - |\vec{B}|$
(E) $\geq |\vec{A}| - |\vec{B}|$

59. The rank of the matrix

$\begin{pmatrix} 2 & 1 & 2 \\ 2 & 1 & 1 \\ 4 & 2 & 2 \end{pmatrix}$

is

(A) 0
(B) 1
(C) 2
(D) 3
(E) 4

60. $\int_1^e \left(1 + \frac{1}{x}\right) dx =$

   (A) e

   (B) $1 + \left(\frac{1}{e}\right)$

   (C) $e - \frac{1}{e^2}$

   (D) $e - 1$

   (E) diverges

61. $\int_0^{\pi/2} \ln[\cos(2x) + 2\sin^2(x)] \, dx =$

   (A) 0

   (B) 1

   (C) -1

   (D) $\ln\left(\frac{\pi}{2}\right)$

   (E) $-\frac{\pi}{2}$

62. $\int_0^{\pi/4} \sin(x)\cos(x) \, dx =$

   (A) 0

   (B) $\frac{1}{4}$

   (C) $\frac{\pi}{2}$

   (D) -1

   (E) $\frac{\sqrt{3}}{2}$

63. $\int_1^{1/x} g'(y)\, dy =$

(A) $g\left(\dfrac{1}{x}\right) - g(1)$

(B) $g\left(\dfrac{1}{x} - 1\right)$

(C) $g'\left(\dfrac{1}{x}\right) - g'(1)$

(D) $\dfrac{1}{g(x)} - 1$

(E) $\ln[g(x)]$

64. $\int_0^{5\pi/2} \cos(x)\, dx =$

(A) -1

(B) 0

(C) 1

(D) $\dfrac{\sqrt{2}}{2}$

(E) $\dfrac{\pi}{2}$

**End of Part B.
Go to Part C.**

PART C

Directions: For each problem, select the best of the choices offered. Every function whose graph appears in this part of the test is to be assumed to have derivatives of all orders at each point of its domain unless otherwise indicated.

Problems 65 and 66 are based on the following information.

$F1$, $F2$, and $F3$ are graphs of the equations

(1) $-2x + y = 2$

(2) $xy = 4$

and

(3) $(x - 4)^2 + (y + 4)^2 = 4$

respectively.

65. Which graph(s) intersect(s) the y axis at finite y?

   (A) $F1$ only
   (B) $F1$ and $F2$ only
   (C) $F1$ and $F3$ only
   (D) all three
   (E) none of them

66. Where does $F1$ intersect $F3$?

   (A) nowhere
   (B) quadrant 1
   (C) quadrant 2
   (D) quadrant 3
   (E) quadrant 4

Problems 67–72 are based on the figure and information below.

f is the function graphed below.

$$g(x) \equiv \int_0^x f(t)\, dt$$

for all $0 < x \leq 12$.

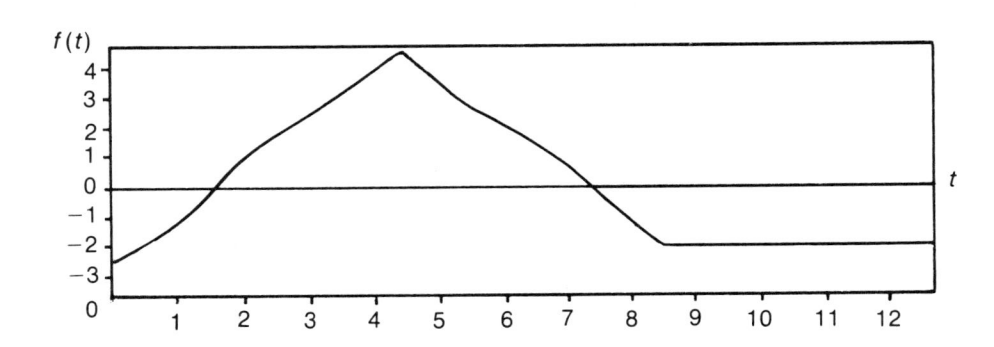

67. $g(0) =$

   (A) -2
   (B) 0
   (C) 0.5
   (D) 1.5
   (E) 3

68. $g(7.5)$ is

   (A) < -4
   (B) < -2 and > -4
   (C) > -1 and < 1
   (D) > 2 and < 6
   (E) > 6

69. Suppose the function $f(t)$ remains horizontal for all large t. The graph of $g(x)$ would show what behavior for large x? Which one of the following would then best describe the behavior of the graph of $g(x)$ for large x?

   (A) it would go as $\dfrac{1}{x}$

   (B) it would grow linearly

   (C) it would decrease linearly

   (D) it would remain constant

   (E) it would go as $\dfrac{1}{x^2}$

70. $g(x) \geq 0$ for all x between (approximately)

   (A) 3 and 11
   (B) 0 and 7.5
   (C) 1.5 and 7.5
   (D) 4 and 7.5
   (E) 7.5 and 12

71. $g(x)$ is an increasing function of x for which of the following intervals?

   (A) 0 to 1.5 and 7.5 to 12
   (B) 1.5 to 7.5 exclusively
   (C) 0 to 1.5 exclusively
   (D) 7.5 to 12 exclusively
   (E) none of the above

72. $g(x)$ is a decreasing function of x in the region

   (A) 0.0 to 1.5 exclusively
   (B) 4.5 to 7.5 exclusively
   (C) 1.5 to 8.5 exclusively
   (D) 0.0 to 1.5 and 9.0 to 12.0 exclusively
   (E) none of the above

Problems 73–77 are based on the following information.

Object A travels at constant speed on the edges of the top face of a cube—clockwise as viewed from above. Object B travels at the same speed on the bottom face of the cube—in the opposite sense. The graph below shows the straight-line distance between A and B as a function of time for $0 \leq t \leq 20$ sec.

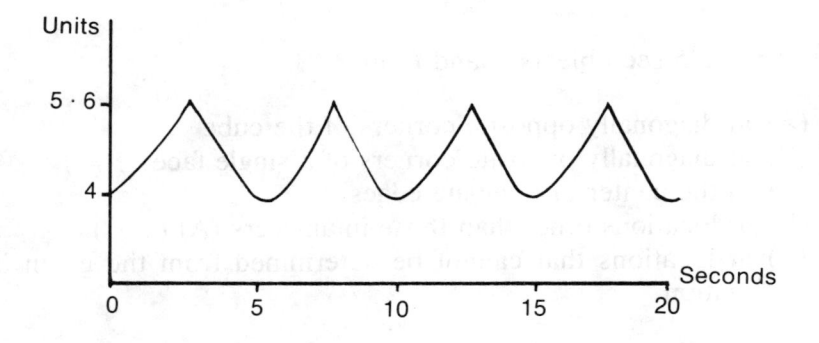

132 / Graduate Record Examination in Engineering

73. If the base of the cube is horizontal, how many times in the interval shown is one object directly above the other?

   (A) 0
   (B) 1
   (C) 2
   (D) 3
   (E) 5

74. Consider a straight line that connects A to B. In the time interval shown, how many times does this line intersect the center point of the cube?

   (A) 0
   (B) 1
   (C) 2
   (D) 3
   (E) 4

75. What is the length of a side of the cube?

   (A) 2
   (B) 4
   (C) 6
   (D) 8
   (E) 12

76. In the time interval shown, how many times does the rate of change of distance between A and B change from positive to negative?

   (A) 0
   (B) 4
   (C) 8
   (D) 16
   (E) none of the above

77. At $t = 2.5$ sec objects A and B are

   (A) at diagonally opposite corners of the cube
   (B) at diagonally opposite corners of a single face
   (C) in the center of opposite edges
   (D) at locations other than those in answers (A) to (C)
   (E) at locations that cannot be determined from the given information

78. If the graph shown below is made antisymmetric by extending it to negative values, which of the following functions best describes the entire graph?

   (A) $y = x^2$
   (B) $y = 1 + \cos(x)$
   (C) $y = x^3$
   (D) $y = x^2 + x + 1$
   (E) $y = \sin(x)$

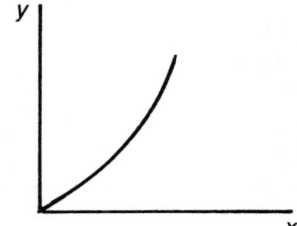

79. Two new scales of temperature, called scale P and scale Q, are so chosen that the boiling point of water is calibrated to read 200° on both scales. A change of 10° on scale P represents the same change in temperature as a change of 20° on scale Q. If follows that

   (A) 0° on scale P corresponds to $-100°$ on scale Q
   (B) 50° on scale P corresponds to $-100°$ on scale Q
   (C) 100° on scale P corresponds to $-200°$ on scale Q
   (D) 0° on scale P represents the melting point of ice
   (E) neither scale P nor scale Q could be an absolute scale of temperature with zero point at absolute zero

80. Each line that is perpendicular to the line $4x - 6y + 5 = 0$ has a slope of

   (A) -2
   (B) $\dfrac{-3}{2}$
   (C) 0
   (D) $\dfrac{2}{3}$
   (E) 1

81. If we consider the interval $(0, \pi)$, the function

   $\sin(x)\cos(x)\tan(x)\csc(x)$

   is

   (A) 1
   (B) 0
   (C) $\cos(x)$
   (D) positive
   (E) none of the above

82. $\lim_{x \to 0^-} \dfrac{e^x - e^{2x}}{\cos(x) - 1} =$

(A) 0
(B) 1
(C) $-\infty$
(D) ∞
(E) e

83. A certain amount of acoustic energy is emitted in a uniform hemispherical pattern from a "point" source. At a distance of 5 m from the source, the intensity is 5 W/m². At a distance of 15 m from the source what is the intensity?

(A) 5 W/m²
(B) 2.5 W/m²
(C) $\dfrac{5}{3}$ W/m²
(D) 1.0 W/m²
(E) $\dfrac{5}{9}$ W/m²

84. The implicit derivative,

$\dfrac{dy}{dx}$

for

$y^2 = x \sin(y)$

is

(A) not determinable
(B) $x \cos(y) + \sin(y)$
(C) $\dfrac{\sin(y)}{2y - x \cos(y)}$
(D) $\dfrac{\sin(y)}{2y}$
(E) $\dfrac{1}{2\sqrt{x \sin(y)}}$

85. In the following equation, ρ represents a fluid density, ω an angular frequency, r a radial distance, and p the fluid pressure.

$$p = p_c + \frac{\rho \omega^2 r^2}{2}$$

What are the primitive SI units of p?

(A) bar
(B) dynes per square centimeter
(C) atmospheres
(D) kg/(m · sec²)
(E) (kg · m)/sec

**End of Part C.
Go to Part D.**

PART D

Directions: Each of the problems or incomplete statements below is followed by five suggested answers or completions. Select the one that is best in each case and then blacken the corresponding space on the answer sheet.

86. All of the statements below except one are classical expectations for the photoelectric effect. Which is a strictly quantum mechanical effect?

    (A) the strength of the incident light is proportional to the photocurrent
    (B) light frequency is independent of the photocurrent
    (C) energy accumulates until an electron is ejected
    (D) there is a photocurrent cutoff frequency for the incident light
    (E) electrons oscillate in response to $\vec{E}$ field of incident light

87. A ball of mass m and velocity v strikes a second ball of equal mass and comes to rest. The time of the interaction is x seconds. The average force experienced by the first ball during this time is

    (A) $2mvx$
    (B) $\dfrac{-mv}{x}$
    (C) $\dfrac{mv^2}{(2x)}$
    (D) mv
    (E) dependent on the detailed nature of the interaction

88. Find the root-mean-square (rms) value of the waveform shown below.

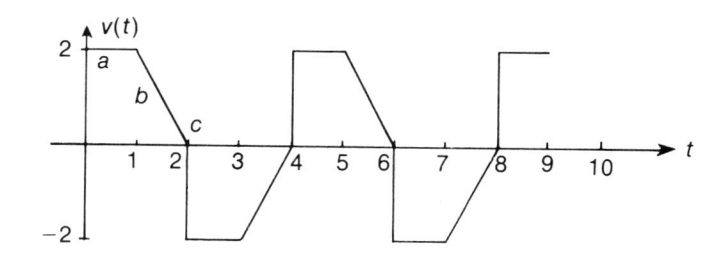

136

(A) 3 V (D) 1
(B) $\frac{8}{3}$ V (E) 0
(C) $2\sqrt{\frac{2}{3}}$ V

89. A model of a submarine is 1/10 the size of the prototype. If the model test is to be conducted in the same fluid as the prototype and the prototype will be running submerged at low velocity of 4 ft/sec, how fast will the model have to move?

   (A) 50 ft/sec (D) 1 ft/sec
   (B) 40 ft/sec (E) 0.8 ft/sec
   (C) 1.2 ft/sec

90. Find the moment of inertia of the area about axis AA for the given rectangle. Assume the given area is homogeneous.

   (A) $\frac{bh^2}{3}$
   (B) $\frac{bh^3}{3}$
   (C) $\frac{bh^3}{6}$
   (D) $\frac{b^3h}{3}$
   (E) $\frac{bh^3}{12}$

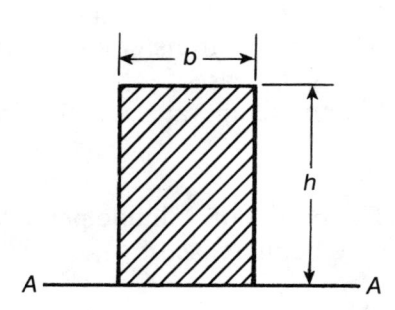

91. A closed container is partially filled with water as shown. Find the gauge pressure at the bottom of the tank if the gauge reads 20 psi and the density of the water is 62.5 lb/ft³.

   (A) 4.34 psi
   (B) 15.66 psi
   (C) 20.00 psi
   (D) 24.34 psi
   (E) 625.00 psi

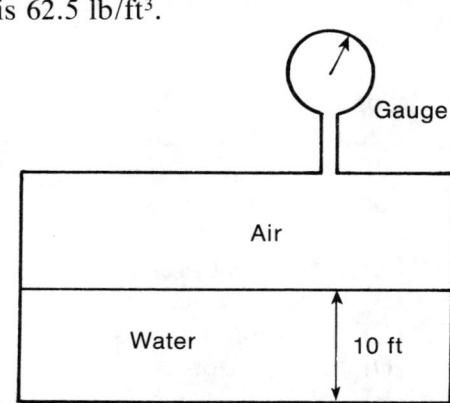

92. In the network shown below, it was determined that the load A consumes 20 W of power with a power factor of 0.8 lagging and load B consumes 16 W with a power factor of 0.6 lagging. It is desired that the power factor of the overall network be 0.9. Determine the value of the resistance R_L.

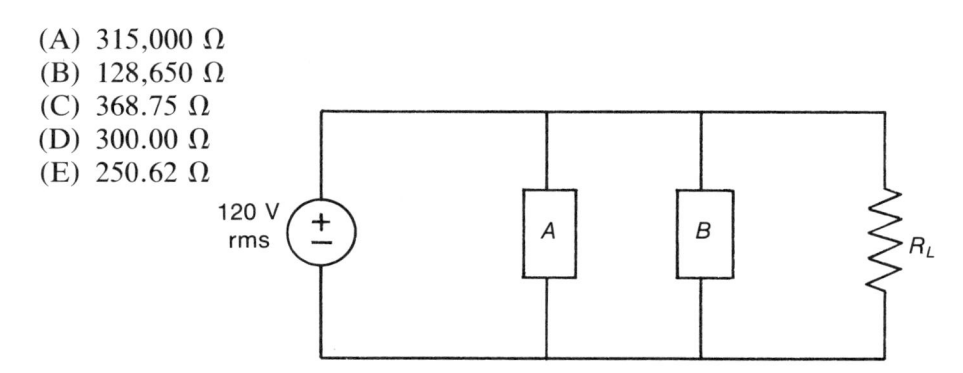

(A) 315,000 Ω
(B) 128,650 Ω
(C) 368.75 Ω
(D) 300.00 Ω
(E) 250.62 Ω

93. A car radiator is provided to increase the rate of heat transfer. Heat transfer is enhanced because

(A) the thermal conductivity is increased
(B) the effective surface area is increased
(C) more material is added to provide more heat conduction
(D) the heat transfer coefficient between the fin and the ambient air is increased
(E) the temperature of the finned heat exchanger is increased

94. The period of a simple pendulum (for very small amplitude motion) may be increased by

(A) increasing the mass of the hanging object
(B) decreasing the mass of the hanging object
(C) increasing the initial displacement from equilibrium
(D) lengthening the string
(E) shortening the string

95. Consider a cylinder fitted with a piston that contains saturated freon-12 vapor at 20°F. Let this vapor be compressed in a reversible adiabatic process until the pressure is 150 lbf/in.² This process is commonly treated as a(an)

(A) isothermal process
(B) isobaric process
(C) isentropic process
(D) throttling process
(E) adiabatic process

96. A body-centered cubic lattice has how many atoms per unit cell?

   (A) 1
   (B) 2
   (C) 3
   (D) 4.5
   (E) 9

97. In the summer, light colored clothing is cooler than dark colored clothing because of differences in their absorptance (absorptivity) α and their emittance (emissivity) ρ. Which one of the following must be true?

   (A) $(\alpha + \rho)$ for dark clothing less than that for light clothing
   (B) $\left(\frac{\alpha}{\rho}\right)$ for dark clothing less than that for light clothing
   (C) $\left(\frac{\alpha}{\rho}\right)$ for dark clothing larger than that for light clothing
   (D) $\alpha = \rho$
   (E) $\alpha = \frac{1}{\rho}$

98. A 4-oz bird sits on a nearly taut clothesline with a 200-lb breaking strength and nearly breaks it. The angular deflection of the line from horizontal is closest to

   (A) 0.01°
   (B) 0.03°
   (C) 0.06°
   (D) 0.23°
   (E) 0.55°

99. To figure out the amount of thermal energy required to bring 20 g of a liquid from 5 degrees below its boiling point to 5 degrees above it, we need some of the following information

   I. the specific heats of the liquid and gas phase
   II. the boiling temperature of the liquid
   III. the liquid's heat of fusion
   IV. the ambient temperature and pressure
   V. the liquid's heat of vaporization

Which of these is required?

(A) I only
(B) I and V
(C) II, III, and IV
(D) I and IV
(E) II, IV, and V

100. Ten kilograms of SO is completely burned to produce SO_2. What is the weight of the SO_2 formed? The atomic weights of sulfur and oxygen are 32 and 16, respectively

(A) 100 N
(B) 120 N
(C) 133 N
(D) 150 N
(E) none of the above

101. A person standing on a moving elevator feels 20% heavier than when at rest. The elevator is

(A) accelerating upward at 2 m/sec²
(B) accelerating upward at 12 m/sec²
(C) accelerating downward at 2 m/sec²
(D) accelerating downward at 6 m/sec²
(E) moving down at constant velocity of 2 m/sec

102. Find the discharge of liquid through pipe C

(A) 0.79 cfs
(B) 6 cfs
(C) 6.28 cfs
(D) 7.06 cfs
(E) 9.42 cfs

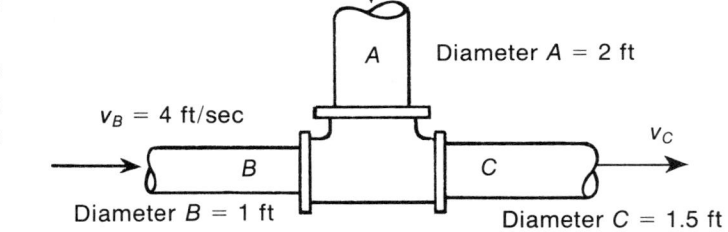

v_A = 2 ft/sec
Diameter A = 2 ft
v_B = 4 ft/sec
Diameter B = 1 ft
v_C
Diameter C = 1.5 ft

103. In a two-component system with four phases present (liquid and gas phases of one component, liquid and solid phase of the other), which of the following may be changed without one or more of the phases disappearing?

(A) T and V
(B) T and P
(C) T alone
(D) P alone
(E) none of these may be changed

104. The graph below is one-half period of a sinusoid. It might represent the time dependence of the

    (A) height of a projectile
    (B) vertical component of a projectile's velocity
    (C) x component of velocity of a particle moving in uniform circular motion
    (D) speed of an object subject to a force that grows linearly in time
    (E) none of the above

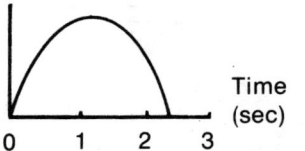

105. The radiation emitted through a small hole in a metallic cavity is

    (A) dependent on the nature of the interior surface
    (B) a function of the temperature only
    (C) a function of both temperature and pressure
    (D) a constant
    (E) dependent on the temperature and on the nature of the surface

106. A tractor cost $120,000. You obtain a loan on the tractor at 11% yearly interest rate with no money down. The first year's payment on this loan is $14,000. How much of this first year's payment goes to pay off the capital cost of the tractor?

    (A) $14,000
    (B) $13,200
    (C) $8,000
    (D) $800
    (E) 0

107. Which of the following functions best describes the time dependence of the thermal output of a piece of copper connected to a constant voltage source (k and k' are constants, ε = temperature coefficient of resistivity)?

    (A) $\dfrac{k}{k' - \varepsilon^2 t}$

    (B) $k(1 + \varepsilon^2 t)$

    (C) $\dfrac{k}{k' + \varepsilon t}$

    (D) $k(1 + \varepsilon t)^2$

    (E) $kt + k'$

108. A 20-kg block resting half on one surface (coefficient of friction $\mu = 0.1$) and half on another surface ($\mu = 0.2$) experiences the time

dependent force $F = (10 + 10t)$ N, where t is in seconds. The block will begin to move when t equals

(A) 0
(B) 1 sec
(C) 2 sec
(D) 2.5 sec
(E) 3.2 sec

109. Given that α, ρ, and τ are the absorptivity (or absorptance), reflectivity (or reflectance), and transmissivity (or transmittance), respectively, of thermal radiation, which one of the following statements is generally true?

(A) $\alpha + \rho \simeq 1$ for most gases
(B) $\alpha + \tau \simeq 1$ for thermally opaque bodies
(C) $\tau \simeq 0$ for most liquids
(D) $\alpha + \rho \simeq 1$ for most solids
(E) $\alpha + \rho + \tau \neq 1$ for all substances

110. A small body at 100°F is placed in a large heating oven whose walls are maintained at 2000°F. The average absorptivity of the body varies with the temperature of the emitter as follows

Temperature (°F)	100°F	1000°F	2000°F
Absorptivity, α	0.8	0.6	0.5

What is the rate at which radiant energy is absorbed by the body per unit surface area? (The Stefan–Boltzmann constant $\sigma = 0.1714 \times 10^{-8}$ Btu/hr-ft²-R^4.)

(A) $2.11 \times 10^{-6} \dfrac{\text{Btu}}{\text{hr-ft}^2}$

(B) $3.38 \times 10^{-6} \dfrac{\text{Btu}}{\text{hr-ft}^2}$

(C) $13.7 \dfrac{\text{Btu}}{\text{hr-ft}^2}$

(D) $3.16 \times 10^5 \dfrac{\text{Btu}}{\text{hr-ft}^2}$

(E) $5.05 \times 10^5 \dfrac{\text{Btu}}{\text{hr-ft}^2}$

111. A cube is resting on very thin supports partly submerged in water as shown. The length of an edge of the cube is l and the cube has a specific gravity of 0.6. Find the minimum rise in the water level, h, that will be required to lift the cube.

    (A) $1\,l$
    (B) $0.7\,l$
    (C) $0.6\,l$
    (D) $0.5\,l$
    (E) $0.4\,l$

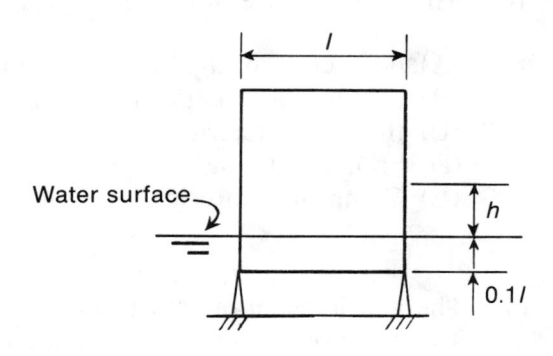

112. For the given velocity (v) equation in a circular pipe of radius r_0 find the maximum velocity.

    $$v = C(r_0^2 - r^2)$$

    where

    r_0 = radius of the pipe
    r = radius measured from the axis of the pipe
    C = constant

    (A) $\frac{2}{3} C r_0$
    (B) $C r_0$
    (C) $\frac{C r_0^2}{2}$
    (D) $C r_0^2$
    (E) 0

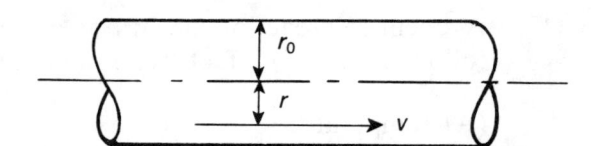

113. A puck on a frictionless table is in pure translation when it strikes a meter stick, causing the latter to undergo both translation and rotation. It appears (incorrectly) that angular momentum is not conserved in the collision. The final angular momentum

    (A) was created during the puck–stick collision
    (B) arose from the friction between the stick and the table
    (C) must be zero, so the puck must end up rotating also

144 / Graduate Record Examination in Engineering

(D) arose from friction between the puck and the stick
(E) is due to none of the above

114. The wave nature of matter is best illustrated by

(A) diffraction of light through an aperture
(B) diffraction of neutrons through a crystal
(C) the photoelectric effect
(D) bremsstrahlung
(E) Compton scattering

115. The acceleration of gravity on the surface of Mars is 12.27 ft/sec². The mass of a man as determined on earth is 175 lbm. What is his weight on Mars?

(A) 14.3 lbf
(B) 66.7 lbf
(C) 175 lbf
(D) 458.9 lbf
(E) 2147.3 lbf

116. Which of the following is *not* true about polymers?

(A) the local growth mechanism is linear
(B) molecular weights can be 1,000,000 or more
(C) the entire molecule is linear
(D) polymers can make plastics and rubbers
(E) polymers are macromolecules and not aggregates

117. A chemical reaction occurs at a certain rate. Which one of the following has *no* effect (with moderate change) on that rate?

(A) temperature
(B) initial concentration of a particular species
(C) size and shape of the reaction vessel
(D) addition of a catalyst (if one exists for the reaction)
(E) presence of electromagnetic radiation

118. For the circuit shown on the following page, determine the direct-current collector-to-emitter voltage, V_{CE}. The values of the components are $\beta = 50$, $V_{CC} = 15$ V, $R_C = 2$ Ω, $R_E = 200$ Ω, $R_1 = 80$ kΩ, $R_2 = 4$ kΩ, $V_{BE} = 0.6$ V.

(A) 15.0 V
(B) 14.1 V
(C) 10.0 V
(D) 0.9 V
(E) 0.6 V

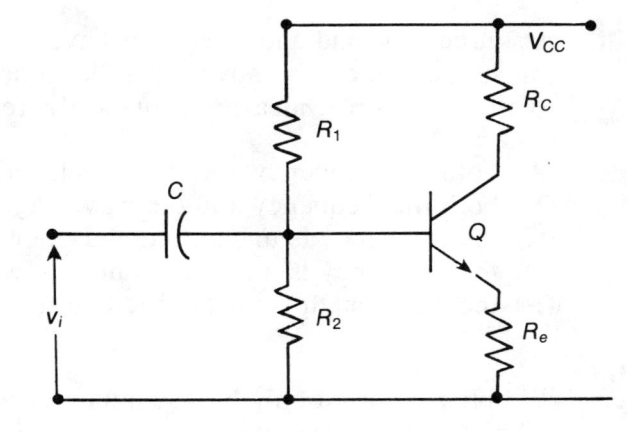

119. Ten cubic meters per second of water enters a rectangular duct as shown in the figure. The upper face of the duct is porous. On the upper face, water leaves at a rate determined linearly by the distance from the end. What is the discharge leaving the duct at point A?

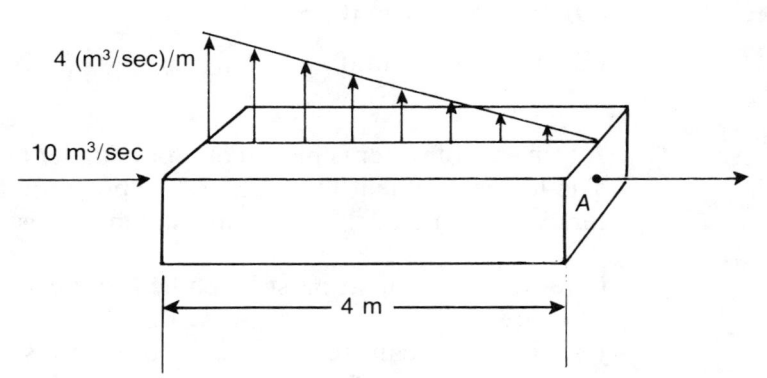

(A) 10 m³/sec
(B) 6 m³/sec
(C) 4 m³/sec
(D) 2 m³/sec
(E) 0 m³/sec

120. The circuit on the left is magnetically coupled to the loop on the right. The induced current in the loop is

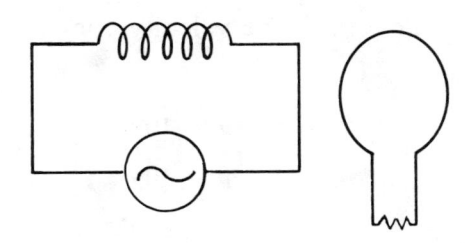

(A) direct current
(B) alternating current of the same frequency as the source
(C) alternating current of twice the frequency of the source
(D) alternating current of half the frequency of the source
(E) zero

121. A source of sound and a receiver have no relative motion, but the transmission medium moves from the source to the receiver. Relative to the source, measurements at the receiver will indicate that

(A) both the frequency and the wavelength are greater
(B) both the frequency and the wavelength are smaller
(C) the frequency is unchanged and the wavelength is smaller
(D) the frequency is unchanged and the wavelength is longer
(E) the wavelength is unchanged and the frequency is smaller

122. A cable supported only by its endpoints assumes the shape shown below. Which of the following functions best describes this curve?

(A) e^x for negative x

(B) $(e^x + e^{-x})$ for x in $(-1, 1)$

(C) $\sin(x)$ for x in $\left(\dfrac{\pi}{2}, \dfrac{3\pi}{2}\right)$

(D) $\tan(x)$ for x in $(0, \pi)$

(E) $\csc(x)$ for x in $(0, \pi)$

123. A thin tray of water is placed outdoors under a clear cloudless night. The lowest ambient temperature is above the freezing point of water. Which one of the following statements is correct?

(A) the water can at most be chilled to the same ambient temperature
(B) the water can freeze because of heat loss by radiation
(C) the water can freeze because of heat loss by convection
(D) the water can freeze because of heat loss by conduction
(E) none of the above

124. For the given flow field, the velocity components are given as follows:

$$u = 10x$$
$$v = 10y$$
$$w = -20z$$

Find the equation of the streamlines in the zx plane.

Note C = constant, u = velocity in the x direction, v = velocity in the y direction, and w = velocity in the z direction.

(A) $z^2x = C$

(B) $x^2z = C$
(C) $yx^2 = C$
(D) $x^{1/2}z = C$
(E) $\ln x = \ln cz$

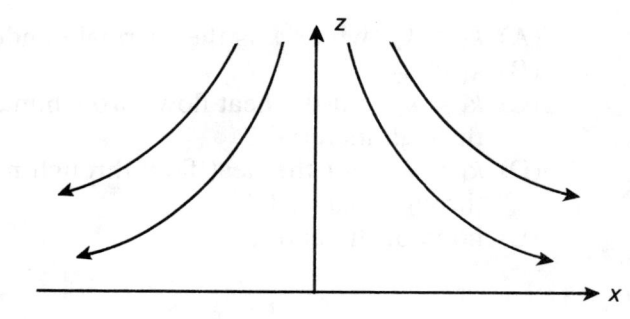

125. At what rate must $10,000 be invested if it is to amount to $13,400 in 6 years compounded annually?

(A) 10%
(B) 6%
(C) 5%
(D) 4%
(E) 0%

126. For the given figure determine the sum of the moments.

(A) 120 lb
(B) 60 lb
(C) 35 lb
(D) 30 lb
(E) 0 lb

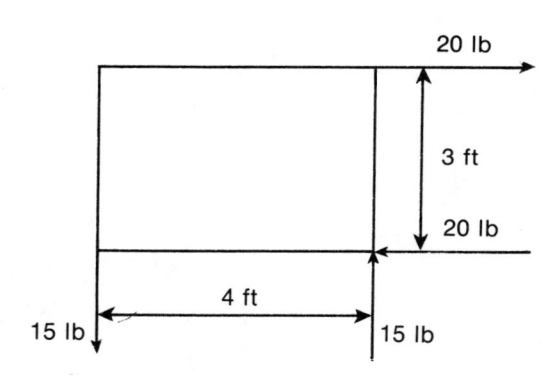

127. The temperature drop through each layer of a two-layer furnace wall is shown in the figure. Assume that the external temperatures T_1 and T_3 are maintained constant and that $T_1 > T_3$. If the thicknesses of the layers, x_1 and x_2, are the same, which one of the following statements is correct?

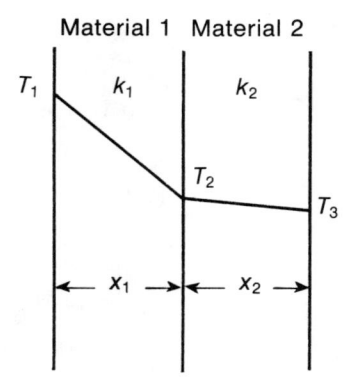

(A) $k_1 > k_2$, where k is the thermal conductivity of the layer
(B) $k_1 < k_2$
(C) $k_1 = k_2$, but the heat flow through material 1 is larger than that through material 2
(D) $k_1 = k_2$, but the heat flow through material 1 is less than that through material 2
(E) none of the above

End of Part D.
Go to Part E.

PART E

Directions: For each problem select the best of the choices offered. Computation and scratch work may be done in this examination book.

128. A region in the xy plane is bounded by the x axis, the y axis, the line $x = c$, and a function $f(x)$ that is always positive. Suppose the area enclosed by this region is found to be proportional to $c * f(c)$ for all $c > 0$ and suppose $f(1) = 1$ and $f(2) = 4$. What is the constant of proportionality?

   (A) e

   (B) 1

   (C) $\frac{1}{2}$

   (D) $\frac{1}{3}$

   (E) $\frac{1}{4}$

129. $\int_a^b f(x)\, dx - \int_a^c f(x)\, dx$

   is equal to

   (A) $\int_a^c f(x)\, dx$

   (B) $\int_b^c f(x)\, dx$

   (C) $\int_c^b f(x)\, dx$

   (D) $-\int_a^c f(x)\, dx$

   (E) none of the above

130. $\int_{-1}^{1} f(x)\, dx + \int_{-1}^{0} g(x)\, dx$ is

   (A) $\int_{-1}^{1} [f(x) + g(x)]\, dx$

149

(B) $\int_{-1}^{0} [f(x) + g(x)] \, dx$

(C) $\int_{0}^{1} [f(x) - g(x)] \, dx$

(D) $2 \int_{-1}^{0} [f(x) + g(x)] \, dx$

(E) not necessarily any of the above

131. $\dfrac{d}{dx} [\cos(x)] =$

   (A) $\sin(x)$
   (B) $-\sin(x)$
   (C) $-\cos(x)$
   (D) $\dfrac{1}{\cos(x)}$
   (E) $\tan(x)$

132. If y is defined implicitly by the equation

   $ye^{xy} = 2$

   then

   $\dfrac{dy}{dx}$

   is

   (A) $ye^{xy} + y$
   (B) e^{xy}
   (C) $\dfrac{-y^2}{(1 + xy)}$
   (D) $xe^{y} + y^2 e^{xy}$
   (E) xye^{y}

133. $\dfrac{d}{dx} [\ln(x) e^x] =$

   (A) $\ln(x) \left(\dfrac{1}{x}\right) e^x$

(B) $\left(\dfrac{1}{x}\right) e^x$

(C) $\left(\dfrac{1}{x} + \ln(x)\right) e^x$

(D) $\left(-\dfrac{1}{x} + \ln(x)\right) e^x$

(E) $\dfrac{1}{x} + \ln(x)$

134. If f is everywhere decreasing (for increasing x), then f' is

   (A) increasing everywhere
   (B) decreasing everywhere
   (C) negative everywhere
   (D) positive everywhere
   (E) none of the above

135. If $f'(x^2)$ denotes

   $$\dfrac{df}{d(x^2)}$$

   and $g'(x)$ denotes

   $$\dfrac{dg}{dx}$$

   then

   $$\dfrac{d}{dx}\left\{\dfrac{f(x^2)}{g(x)}\right\}$$

   (A) $\dfrac{[f(x^2)g'(x) - g(x)f'(x^2)]}{g^2(x)}$

   (B) $\dfrac{[g(x)f'(x^2)2x - f(x^2)g'(x)]}{g^2(x)}$

   (C) $\dfrac{f'(x^2)2x}{g'(x)}$

   (D) $\dfrac{f'(x^2)2x}{(g')^2(x)}$

   (E) none of the above

136. The Taylor series expansion of e^x around zero is

(A) $\sum_{1}^{\infty} \frac{x^n}{n!}$ for all n

(B) $\sum_{0}^{\infty} \frac{x^n}{n!}$ for all n

(C) $\sum_{1}^{\infty} \frac{x^n}{n!}$ for odd n

(D) $\sum_{0}^{\infty} \frac{x^n}{n!}$ for even n

(E) $\sum_{1}^{\infty} \frac{x^n}{n!}$ for even n

137. $\lim_{x \to 0} \frac{e^x - x - 1}{x^2} =$

(A) 0

(B) $\frac{1}{2}$

(C) 1

(D) 2

(E) ∞

138. $\sum_{k=0}^{\infty} \frac{-\left(\frac{-\pi}{2}\right)^{2k+1}}{(2k + 1!)} =$

(A) 0

(B) 1

(C) $\frac{\pi}{2}$

(D) π

(E) none of the above

Problems 139 and 140 are based on the following information:

Box A has three orange balls and two red balls.
Box B has two orange balls and four red balls.

139. If two balls are selected randomly (without replacement) from box A, what is the probability that at least one is red?

   (A) 12%
   (B) 30%
   (C) 42%
   (D) 60%
   (E) 70%

140. If two balls are selected randomly (without replacement) from A and two more balls are selected randomly from B, what is the probability that all four balls are orange?

   (A) 12%
   (B) 10%
   (C) 4%
   (D) 2%
   (E) 1.2%

End of Test 2.

Sample Test 2
Answer Key

1.	D	36.	A	71.	B	106.	D
2.	D	37.	D	72.	E	107.	C
3.	B	38.	C	73.	E	108.	C
4.	B	39.	A	74.	E	109.	D
5.	E	40.	A	75.	B	110.	D
6.	A	41.	B	76.	B	111.	D
7.	B	42.	C	77.	C	112.	D
8.	A	43.	B	78.	C	113.	E
9.	D	44.	B	79.	B	114.	B
10.	C	45.	B	80.	B	115.	B
11.	E	46.	C	81.	D	116.	C
12.	A	47.	C	82.	D	117.	C
13.	D	48.	C	83.	E	118.	B
14.	B	49.	C	84.	C	119.	D
15.	B	50.	C	85.	D	120.	B
16.	C	51.	B	86.	D	121.	D
17.	D	52.	C	87.	B	122.	B
18.	C	53.	D	88.	C	123.	B
19.	C	54.	C	89.	B	124.	B
20.	C	55.	B	90.	B	125.	C
21.	E	56.	B	91.	D	126.	E
22.	B	57.	C	92.	C	127.	B
23.	B	58.	C	93.	B	128.	D
24.	E	59.	C	94.	D	129.	C
25.	D	60.	A	95.	C	130.	E
26.	B	61.	A	96.	B	131.	B
27.	A	62.	B	97.	C	132.	C
28.	C	63.	A	98.	B	133.	C
29.	A	64.	C	99.	B	134.	C
30.	B	65.	A	100.	C	135.	B
31.	B	66.	A	101.	A	136.	B
32.	C	67.	B	102.	E	137.	B
33.	D	68.	E	103.	E	138.	B
34.	E	69.	C	104.	C	139.	E
35.	C	70.	A	105.	B	140.	D

Sample Test 2
Solutions

PART A

1. Correct answer: **(D)**
   If the connection in series is aiding, then the equivalent inductance is given by
   $$L_{\text{aid}} = L_1 + L_2 + 2M$$
   When the connection is opposing
   $$L_{\text{opp}} = L_1 + L_2 - 2M$$
   Hence,
   $$M = \frac{1}{4}(L_{\text{aid}} - L_{\text{opp}}) = \frac{1}{4}(20 - 12) = 2 \text{ H}$$

2. Correct answer: **(D)**
   $$W = \text{weight} - \text{buoyancy}$$
   $$\text{Buoyancy} = \gamma \times \text{volume} = 70 \text{ lb/ft}^3 \times 0.1 \text{ ft}^3 = 7 \text{ lb}$$
   Therefore,
   $$W = 10 - 7 = 3 \text{ lb}$$

3. Correct answer: **(B)**
   The relationship among $V_1, V_2, I_1, I_2, n_1, n_2$ for an ideal transformer is given by
   $$\frac{n_1}{n_2} = \frac{V_1}{V_2} = \frac{I_2}{I_1}$$
   Hence,
   $$\frac{n_1}{n_2} = \frac{I_2}{I_1} = \frac{20}{0.2} = \frac{100}{1}$$

4. Correct answer: **(B)**
   The temperature of a gas is a measure of the rms speed of the component molecules. A common derivation involves the experimental observation that
   $$PV = nRT$$

155

and the derivation from Newton's Second Law that

$$PV = \frac{1}{3} nM\, v_{rms}^2$$

resulting in

$$T \propto v_{rms}^2$$

Doubling the temperature raises the rms speed by a factor of $\sqrt{2}$. Alternately, the equipartition of energy theorem says that each degree of freedom in a system gets $1/2\, kT$ of energy. For an ideal monatomic gas there are three translational degrees of freedom, so

$$\frac{1}{2} Mv^2 = \frac{3}{2} kT$$

Again, $v \propto \sqrt{T}$.

5. Correct answer: **(E)**
   The net torque $\vec{\tau}_n$ on a system is equal to the time rate of change of the system's angular momentum

   $$\vec{\tau}_n = \frac{d\vec{L}}{dt}$$

   A constant $\vec{L}$, therefore, implies $\vec{\tau}_n = 0$. This condition may be satisfied if there are torques present, as long as there is no net torque.

6. Correct answer: **(A)**
   Pedaling drives the back wheel directly (in most bicycles) and since this rolls without slipping, we may ignore the front wheel. If we pedal at p revolutions per minute (rpm) and the wheel diameter is d, then our linear velocity is $p * \pi * d$ m/min. In other words, v is directly proportional to d and so an increase of d to d'' means the velocity increases by a factor of

   $$\frac{d''}{d}$$

7. Correct answer: **(B)**
   Light transmitted into an optically denser medium bends toward the surface normal in accordance with the law of refraction,

   $$n_1 \sin \theta_1 = n_2 \sin \theta_2$$

Here the angles are small so we may approximate $\sin\theta$ as θ, so

$$\theta_2 \approx \theta_1 \frac{n_1}{n_2} = 1°\left(\frac{1}{1.5}\right) = 0.67°$$

This is a 0.33° deviation toward the surface normal.

8. **Correct answer: (A)**
   The correct statement is (A) for a stable system.

9. **Correct answer: (D)**
   The force on the mass (in the direction of its motion) is $-kx$ so the equation of motion is $-kx = ma$ or

   $$\frac{d^2x}{dt^2} = -\frac{k}{m}x$$

   If we look for an oscillatory solution of the form $x = A\cos(\omega t)$ then we find

   $$\omega^2 = \frac{k}{m} \text{ or } \omega = 2\pi f = \sqrt{\frac{k}{m}}$$

   $$k = 4\pi^2 f^2 m = 4\pi^2(0.5)^2 4 = 4\pi^2 \frac{N}{m}$$

10. **Correct answer: (C)**
    The mechanical work done during expansion is

    $$W = \int_{V_1}^{V_2} P\, dV$$

    From the ideal gas law,

    $$PV = nRT, \ P = \frac{nRT}{V}$$

    so

    $$W = \int_{V_1}^{V_2} nRT\, \frac{dV}{V}$$

    Since the process is isothermal, T comes outside the integral and

    $$W = nRT \int_{V_1}^{V_2} \frac{dV}{V} = nRT \ln \frac{V_2}{V_1}$$

158 / Graduate Record Examination in Engineering

We have made no use of the pressure of the gas in the cylinder. Had we taken the gas alone as our system, we would have needed the internal pressure exerted on the interior piston face as a function of V.

11. Correct answer: **(E)**
Apply Kirchhoff's Current Law at the node indicated by dashed lines

$$50 - i_x - 20 + 4i_x = 0$$
$$i_x = -10 \text{ mA}$$

Hence,

$$v_x = 10K \times i_x$$
$$v_x = -100 \text{ V}$$

12. Correct answer: **(A)**
One of the postulates of special relativity is that all inertial observers measure the same value for the speed of light. A sphere whose radius grows at the speed of light is therefore an invariant object in inertial frames. Blocking off half of the sphere to make a hemisphere has no effect on the problem.

13. Correct answer: **(D)**
Conservation of energy is used to solve this problem. The electrostatic potential is the electric energy per unit charge, or $U = qV$. The only other form of energy present here is kinetic energy.

$$qV_1 + \frac{1}{2} mv_1^2 = qV_2 + \frac{1}{2} mv_2^2$$

$$v_2 = \sqrt{v_1^2 + \frac{2q\Delta V}{m}}$$

Here,

$$v_1 = 10^7 \text{ m/sec},$$
$$\Delta V = 1200 \text{ V}$$

and

$$\frac{q}{m} = \frac{1.6 \times 10^{-19} \text{ C}}{9.11 \times 10^{-31} \text{ kg}} \approx 2 \times 10^{11} \frac{\text{C}}{\text{kg}}$$

$$\frac{2q \, \Delta V}{m} \approx 4 \times 10^{14} \frac{\text{CV}}{\text{kg}} = 4 \times 10^{14} \frac{\text{m}^2}{\text{sec}^2}.$$

$$V_2 \approx \sqrt{5 \times 10^{14} \frac{\text{m}^2}{\text{sec}^2}} \approx 2.2 \times 10^7 \frac{\text{m}}{\text{sec}}$$

Since v is still less than 10% of the speed of light, we expect less than 5% error due to the use of classical mechanics.

14. Correct answer: **(B)**
    The acronym "laser" stands for *l*ight *a*mplification by *s*timulated *e*mission of *r*adiation. A laser works by heavily populating a single atomic state and introducing a single "trigger" photon that starts a cascade of transitions to a lower energy state. Each stimulated photon has the same energy and phase as the original photon.

15. Correct answer: **(B)**
    At the critical point, the liquid changes to the vapor phase without undergoing a constant-temperature vaporization process. The temperature, pressure, and specific volume at the critical point are called the critical temperature, critical pressure, and critical volume, respectively.
    At a constant pressure greater than the critical pressure, there is no definite change in phase from liquid to vapor and no definite point at which there is a change from the liquid phase to the vapor phase. At a constant pressure lower than the critical point, there is a constant-temperature vaporization process, wherein the liquid phase and vapor phase coexist.

16. Correct answer: **(C)**
    A diverging lens may produce a perfectly sharp image even though it is on the object side of the lens. Each of the remaining choices is easily ruled out. The focal length of a lens depends on the index of refraction of glass, which varies with the wavelength of the light used. Thus, white light will result in a different focal point for each component frequency. If the image is being projected on a screen at the wrong point in space, converging rays will fail to meet. The thin lens equation, which guarantees a sharp image point for each object point, is based on the paraxial approximation. Finally, an object as large as a lens will violate the paraxial approximation of small angles.

160 / Graduate Record Examination in Engineering

17. Correct answer: **(D)**
Remember that

$$i_L(0^-) = i_L(0^+)$$

and

$$v_c(0^-) = v_c(0^+)$$

For $t < 0$ the circuit is at steady state and the inductor acts as a short circuit while the capacitor, after it is charged fully, acts as an open circuit. Hence, at $t = 0^-$ the circuit reduces to

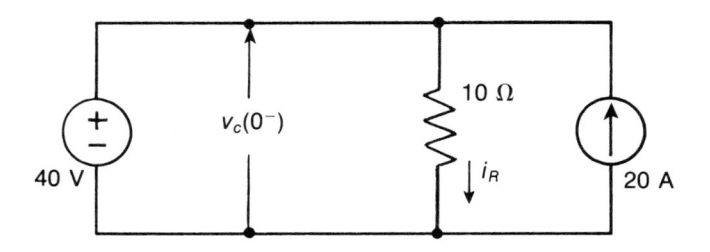

Thus, using Kirchhoff's Voltage Law and Kirchhoff's Current Law

$$40 = [i_L(0^-) + 20] \cdot 10$$

$$i_L(0^-) = i_L(0^+) = -16 \text{ A}$$

In the above circuit determine $v_c(0^+)$. [Answer: $v_c(0^-) = v_c(0^+) = 40$ V.]

18. Correct answer: **(C)**
With a 30°C or 40°C difference in temperature between the person and his or her environment, heat loss by conduction may be significant. In a windy environment, convection will also be significant. Radiation (of the blackbody variety) is only significant at high temperatures (1000°C). Actually, there may be a fair amount of energy loss from a person due to radiation, but his or her surroundings are likely transferring almost as much energy back to the body through this process.

19. Correct answer: **(C)**
Quantum number n, the principal quantum number, determines the mean orbital size or radius. Quantum number l, the orbital angular momentum quantum number, determines the eccentricity of the orbit. $l = 0$ states pass through the nucleus, indicating no orbital angular momentum, maximum l states correspond to circular orbits and therefore maximum angular momentum. m_l and m_s have to do with the projections of orbital and spin angular momentum vectors along the z axis and are therefore related to the three-dimensional

orientation of the orbital plane and the electron's intrinsic spin vector.

20. **Correct answer: (C)**
Write *KCL* at $t = 0$.

$$C\frac{dv}{dt} + \frac{v}{R} = 0$$

Solution of this equation yields [observe that at $t = 0$ $v(0) = E_0$]

$$v = E_0 \exp\left(-\frac{t}{RC}\right)$$

at $t = 0.1$ sec.

$$l = E_0 \exp\left(-\frac{0.1}{20 \text{ k}\Omega \cdot 20 \text{ }\mu\text{F}}\right)$$
$$E_0 = 1.28 \text{ V}$$

21. **Correct answer: (E)**
The Lorentz force on a test charge q is $\vec{F} = q(\vec{E} + \vec{v} \times \vec{B})$. The initial force is due to the electric field because $\vec{v} = 0$. The resulting motion is along the wire. Now $(\vec{v} \times \vec{B})$ makes a contribution. $\vec{B}$ is azimuthally around the wire so $(\vec{v} \times \vec{B})$ points radially inward from the right-hand rule. Now the charge moves radially into the wire and along it. $(\vec{v} \times \vec{B})$ for this new velocity only has components pointing in these directions; there can never by any azimuthal motion.

22. **Correct answer: (B)**
In this process, work has been done on the system, thus $W < 0$. Since the container is thermally insulated, $Q = 0$. Thus, according to the First Law of Thermodynamics

$$Q - W = \Delta U$$

The fact that $W < 0$, $Q = 0$ give rise to $\Delta U > 0$, that is, increasing the internal energy of the system.

23. **Correct answer: (B)**
Voltage across L_1:

$$V_1 = L_1 \frac{di}{dt} - M\frac{di}{dt}$$

Voltage across L_2:

$$V_2 = -M\frac{di}{dt} - L_2\frac{di}{dt}$$

$$V_{xy} = V_1 + V_2 = (L_1 + L_2 - 2M)\frac{di}{dt}$$

Hence,

$$L_{eq} = L_1 + L_2 - 2M = 4 + 6 - 2 = 8\ H$$

Solve the above problems with dots at different locations in L_1 and L_2.

24. Correct answer: **(E)**
Since $x = 0$ at $t = 0$ and $x = R$ at $t = 2$ sec there were some whole number of turns (perhaps none) and one leftover quarter turn. We have no information as to how many complete turns were made so ω is indeterminate.

25. Correct answer: **(D)**

$$v \equiv \sqrt{\left(\frac{dx}{dt}\right)^2 + \left(\frac{dy}{dt}\right)^2 + \left(\frac{dz}{dt}\right)^2}$$

Here

$$\frac{dx}{dt} = -R\omega\sin(\omega t + \phi)$$

$$\frac{dy}{dt} = R\omega\cos(\omega t + \phi)$$

and

$$\frac{dz}{dt} = k$$

Putting these in the expression for v and using

$$\sin^2\theta + \cos^2\theta = 1$$

we find v as a function of k, R, and ω.

26. Correct answer: **(B)**
During the process, heat is transferred from the mixture to the surrounding water and no work is done. The change in internal

energy of the system is negative since

$$Q - W = \Delta U$$
$$W = 0, Q < 0 \text{ and } \Delta U < 0$$

27. **Correct answer: (A)**
This is a problem in the vector addition of velocities. The observed velocity is the sum of the swimmer's motion and the motion of the medium. A sketch is important here.

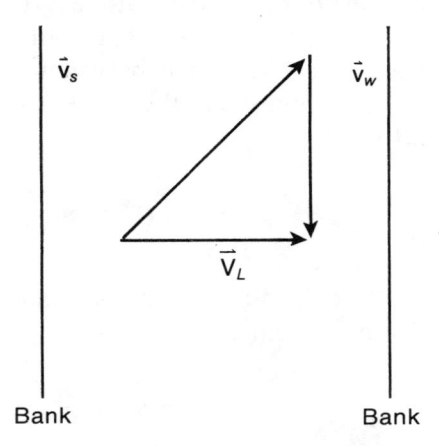

28. **Correct answer: (C)**
The dipole field is easily computed for a point along the dipole axis as the sum of two point charge fields

$$+\frac{kq}{r^2} - \frac{kq}{(r+d)^2} = \frac{kq}{r^2}\left(1 - \frac{1}{(1+d/r)^2}\right) \simeq \frac{2kqd}{r^3}$$

with the usual approximations for

$$\frac{1}{(1+x)^2}$$

for small $x = \dfrac{d}{r}$

that is, large r. The large distance dependence of field strength is $1/r^3$. Other choices can be ruled out. Each field line starting on a positive charge will end on the negative charge. The field must be symmetric about the dipole axis and sufficiently close to either charge, its $1/r^2$ singularity must mask the field of the other charge.

29. **Correct answer: (A)**
The cooling process has two stages. First, steam at 100°C cools to liquid water at 100°C. The heat released in this condensation is (540

cal/g)(50 g) = 27,000 cal or 27 kcal. In the second stage, water at 100°C is cooled to water at 80°C. The heat released is

$$\Delta q = mc\Delta T = (50 \text{ g})(1 \text{ cal/g C°})(20\text{C°}) = 1000 \text{ cal}$$

In total 28,000 cal = 28 kcal of heat are transferred.

30. Correct answer: **(B)**
A two-slit pattern is simply alternate constructive and destructive interference as the difference in path length takes on integer and odd half-integer wavelength values. The path length difference is $a * \sin(\theta)$, where a is the distance between slits and θ is the angular position of interest relative to the center of the pattern. Bright spots will be found when

$$a \cdot \sin(\theta) = 2n \left(\frac{\lambda}{2}\right)$$

Dark spots will be found when $2n$ is replaced by $2n + 1$. From the geometry of the experiment,

$$\sin(\theta) = \frac{x}{d}$$

where x is the linear distance of the point of interest from the center of the pattern and d is the distance to the screen from the slits. Putting these equations together, we find that the distance between bright and dark zones is a constant. Pattern D is a single slit diffraction pattern that always modulates the two-slit pattern, but you were told to ignore this effect.

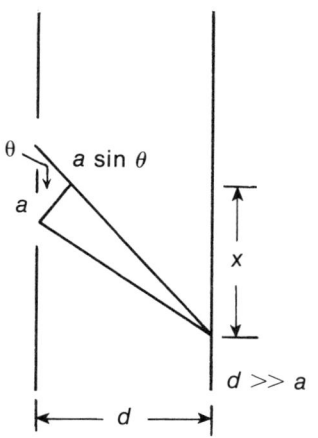

31. Correct answer: **(B)**
From the free body

$$\sum \text{Moment at } A$$
$$(10 \text{ ft})F_D = (20 \text{ lb})(5 \text{ ft})$$

Therefore,

$$F_D = 10 \text{ lb}$$

$$\sum \text{Vertical force}$$

$$F_A + F_D = 20 \text{ lb}$$

Therefore,

$$F_A = 20 - 10 = 10 \text{ lb}$$

Taking a free body at joint A

$$\sum \text{Vertical forces}$$

$$F_A = F_{AB} \sin 45° = F_{AB} \frac{1}{\sqrt{2}}$$

$$F_{AB} = 10 \times \sqrt{2} = 14.1 \text{ lb}$$

32. **Correct answer: (C)**
Refer to the figure in Problem 31.

$$\sum \text{Horizontal forces}$$

$$F_{AC} = F_{AB} \cos 45° = F_{AB} \frac{1}{\sqrt{2}}$$

$$F_{AB} = 14.1 \frac{1}{\sqrt{2}} = 10.0 \text{ lb}$$

33. **Correct answer: (D)**
The First Law of Thermodynamics states that for a closed system the net energy transferred as heat Q and as work W is equal to the change in internal energy, U; that is,

$$Q - W = \Delta U$$

Note that conventionally, heat transferred into the system and work done by the system are taken to be positive.

34. **Correct answer: (E)**
The forces present in the nucleus are the proton–proton coulomb repulsion and the nuclear (attractive) strong force that is independent of the type of nucleon (the n–n force is the same as the n–p force or the p–p force). As the number of protons grows, the coulomb repulsion grows; therefore, extra attractive forces are needed to keep the nucleus stable. Adding neutrons provides this stability.

166 / Graduate Record Examination in Engineering

35. Correct answer: **(C)**
The hydrostatic head is measured perpendicular to the velocity in this problem and is H at the mouth of the Pitot tube. When the tube is placed in the stream, the water will move into and up the tube until the dynamic head is converted into a potential head Δh.

36. Correct answer: **(A)**
The activity due to a single decay process obeys the law

$$A = A_0 \exp(-\lambda * t)$$

where λ is the decay constant. Here we have two such processes with equal constants A_0. We then have

$$\frac{A_1}{\exp(-\lambda_1 \cdot t)} = \frac{A_2}{\exp(-\lambda_2 * t)}$$

or

$$\frac{A_1}{A_2} = \exp[(\lambda_2 - \lambda_1) * t]$$

Since this ratio is e for $t = 1$ hr, we have $(\lambda_2 - \lambda_1) = 1$ hr^{-1}, or the difference in decay rates is 1 hr^{-1}.

37. Correct answer: **(D)**
Summing moments about A

$$P \times 8 = 5 \times 10$$

therefore,

$$P = \frac{50}{8}$$

$$P = \text{weight cube} - \text{buoyant upward force}$$
$$= 112.4 \text{ lb/ft}^3 \times a^3 - 62.4 \text{ lb/ft}^3 \times a^3$$

where

$$a = \text{length of the side of the cube}$$

Therefore,

$$P = (112.4 - 62.4)a^3 = \frac{50}{8}$$

and

$$a^3 = \frac{1}{8},$$

and

$$a = \frac{1}{2} \text{ ft}$$

38. **Correct answer: (C)**
    The self-inductance L of a coil is defined as the ratio of the magnetic flux ϕ passing through its N turns to the current inducing that flux, I, or

    $$L = \frac{N\phi}{I}$$

    For a coil, Ampère's law tells us that $B = \mu_0 nI$, where μ_0 is the magnetic permeability of free space (modified if there is a core), n is the number of turns per unit length N/l, and I is the current through the coil. Since $\phi = BA$, we find

    $$L = \mu_0 N^2 \frac{A}{l}$$

39. **Correct answer: (A)**
    This problem is solved by the linear equation

    $$x(70.92) + (1 - x)(68.93) = 69.724$$

    x is 0.399, or 39.9%.

40. **Correct answer: (A)**
    Draw the circuit at steady state and at $t = 0^+$ (i.e., right after the switch opens). At a steady state, capacitors will be charged to

    $$v_{c_1}(0^-) = \frac{24}{3 + 9 + 6} \times 9 = 12 \text{ V}$$

    $$v_{c_2}(0^-) = \frac{24}{3 + 9 + 6} \times 6 = 8 \text{ V}$$

and the circuit will look like

and

$$i(0^-) = \frac{24}{3 + 9 + 6} = \frac{4}{3} \text{ A}$$

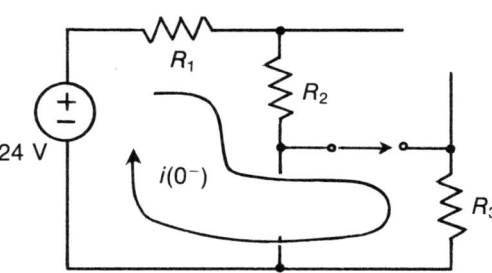

At $t = 0^+$, the circuit will look like

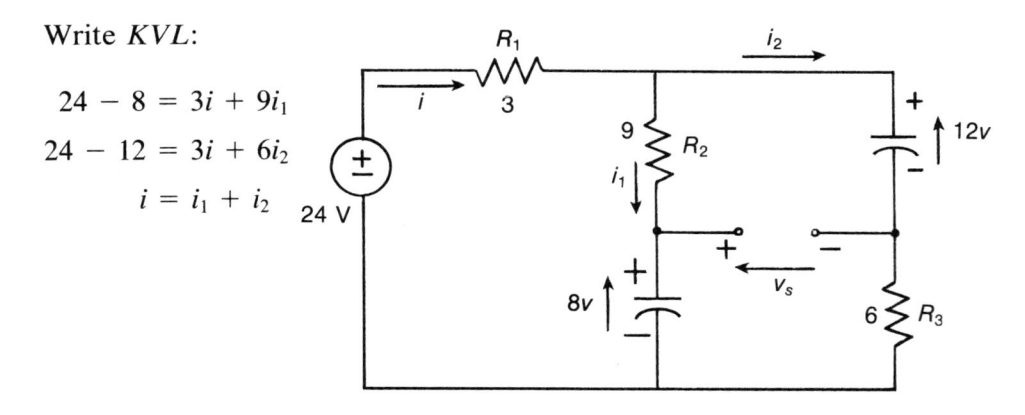

Write KVL:

$$24 - 8 = 3i + 9i_1$$
$$24 - 12 = 3i + 6i_2$$
$$i = i_1 + i_2$$

Solving simultaneously yields

$$i_1 = \frac{36}{33}, \quad i_2 = \frac{32}{33}$$

Using KVL again,

$$8 - v_s(0^+) = 6 \times i_2$$

or

$$v_s(0^+) = \frac{24}{11} \text{ V}$$

(Can you determine $V_s(\infty)$? [Answer: $V_s(\infty) = 24$ V.]

41. Correct answer: **(B)**
    Assume the sphere is rotating about the z axis through its center. The angular velocities of all points on the sphere are identical. The linear velocity of each point is given by its linear distance traveled per unit time, which is

$$(2\pi r/\text{revolution}) * (\text{revolutions}/\text{sec})$$

or

$$2\pi f r = \omega r$$

where r is the effective radius for the point—its distance to the axis of rotation. Thus, the linear velocity is proportional to r and $r = R \cos(\theta)$, where R is the radius of the sphere and θ is the angle up from the equator, that is, the latitude.

42. Correct answer: **(C)**
The support should be located at the center of mass of the system.

$$x_{cm} = \frac{\sum_{i=1}^{n} m_i x_i}{\sum_{i=1}^{n} m_i}$$

where there are n masses m_i located at positions x_i. Evaluating the center of mass for this system, we find

$$x_{cm} = \frac{(1)(1) + (2)(2) + (3)(3) + (4)(4)}{(1 + 2 + 3 + 4)} = 30 \text{ kg} * \text{m}/10 \text{ kg} = 3 \text{ m}$$

43. Correct answer: **(B)**
Transition elements are defined by their tendency to fill d subshells while s subshells with larger principal quantum numbers are already at least partially occupied. Transition elements could not be nearly as inert as the noble gases since they have one or two electrons in their outer shells. Since their outer shells differ, we do not expect all transition elements to be chemically indistinguishable. Finally, with uncompleted shells, we do not expect their angular momentum vectors to cancel.

44. Correct answer: **(B)**
Sodium is deficient in one electron and chlorine has a surplus of one electron. A stable configuration is possible if there is an outright transfer of the electron. Such bonds are called ionic or electrovalent.

45. Correct answer: **(B)**
Compounding money semiannually for 3 years at 10% is the same as compounding the money at 5% every 6 months or

$$\$1,000 \times (1 + 0.05)^6 = \$1,340$$

PART B

46. Correct answer: **(C)**

$$G\left(\frac{1}{x}\right) = \frac{1/x}{(1/x) + 2}$$

Multiply the numerator and the denominator by x:

$$G\left(\frac{1}{x}\right) = \frac{1}{1 + 2x}$$

47. Correct answer: **(C)**
A nonzero matrix has a rank R if at least one of its R-square minors is different from zero. If the $(R + 1)$ square minor exists, it is equal to zero.

First calculate the determinant (D) of the given 3×3 matrix. Place the first two rows outside and to the right of the given matrix:

$$D = \begin{vmatrix} 1 & 2 & 1 \\ 3 & 3 & 0 \\ 1 & 1 & 0 \end{vmatrix} \begin{matrix} 1 & 2 \\ 3 & 3 \\ 1 & 1 \end{matrix}$$

The procedure to calculate the value of a determinant is to multiply all the elements in a diagonal starting from upper left proceeding to lower right. These terms are positive. Starting from upper right going to lower left, the terms are negative. Therefore,

$$D = (1 \times 3 \times 0) + (2 \times 0 \times 1) + (1 \times 3 \times 1) - (1 \times 3 \times 1)$$
$$- (1 \times 0 \times 1) - (2 \times 3 \times 0) = +3 - 3 = 0$$

Finding D of the lower order matrix

$$\begin{vmatrix} 1 & 2 \\ 3 & 3 \end{vmatrix} = (1 \times 3) - (2 \times 3) = 3 - 6 = -3$$

Therefore,

$$R = 2$$

48. Correct answer: **(C)**
Let $u = 3x$ and $du = 3\,dx$

Therefore,

$$\int \sin 3x \, dx = \frac{1}{3} \int \sin u \, du$$

$$\int \sin u \, du = -\cos u$$

$$\frac{1}{3} \int \sin u \, du = -\frac{1}{3} \cos u$$

$$\int_0^{\pi/12} \sin 3x \, dx = -\frac{1}{3} \cos 3x \Big]_0^{\pi/12}$$

$$= -\frac{1}{3} \cos \left(3 \frac{\pi}{12}\right) + \frac{1}{3} \cos (0)$$

$$= -\frac{1}{3} \cos \left(\frac{\pi}{4}\right) + \frac{1}{3}$$

$$= -\frac{1}{3} \left(\frac{1}{\sqrt{2}}\right) + \frac{1}{3} = \frac{1 - \frac{1}{\sqrt{2}}}{3}$$

49. Correct answer: **(C)**

$$y = \ln(2x + 3)$$

Let

$$u = 2x + 3$$

then

$$du = 2 \, dx$$
$$y = \ln(2x + 3)$$
$$\frac{dy}{du} = \frac{1}{u}$$

and

$$\frac{dy}{2dx} = \frac{1}{2x + 3}$$

Therefore,

$$\frac{dy}{dx} = \frac{2}{2x + 3}$$

50. Correct answer: **(C)**

Let $u = e^x$ and $v = \sin x$

Therefore,

$$y = uv$$

The rule is

$$dy = du\, v + u\, dv$$
$$dy = e^x \sin x\, dx + e^x \cos x\, dx$$
$$\frac{dy}{dx} = e^x(\sin x + \cos x)$$

51. Correct answer: **(B)**

$$\int_a^b g'(t)dt = \int_a^b \frac{dg(t)}{dt} dt = \int_a^b dg(t) = g(t)\Big]_a^b = g(b) - g(a)$$

With the limits $a = 0$ and $b = \dfrac{1}{x}$

$$= g\left(\frac{1}{x}\right) - g(0)$$

52. Correct answer: **(C)**

For F to be a cumulative distribution function, we require $f(x) \geq 0$ and a total probability of one. This means that $F(\infty) = F(1)$ must equal 1, or

$$c \int_{-1}^{1} x^2\, dx = 1$$

The indefinite integral is

$$\frac{cx^3}{3}$$

so the definite integral is $2c/3$. c must equal $3/2$.

53. Correct answer: **(D)**

Replacing k with $+k^2$, the solutions are exponential functions—which in fact satisfy the first-order equation $y' = ky$. Replacing k with $-k^2$, we obtain sinusoidal solutions. The only other solution is the trivial one $y(x) = 0$.

54. **Correct answer: (C)**
    A necessary condition for a matrix to have an inverse is that its determinant be nonzero. A quick calculation of each determinant reveals matrix (C) to be singular. Alternately, linear independence of the rows of the matrix is necessary and sufficient for an inverse. In matrix (C), the rows are not linearly independent since row 2 is a scalar multiple of row 1.

55. **Correct answer: (B)**
    Any n by m matrix may be multiplied by any m by n matrix forming an n by n product matrix. In the most common case, n is equal to m. The formula for matrix multiplication is

    $$C_{ij} = \sum_k A_{ik} B_{kj}$$

    That is, rows on the left are multiplied by columns on the right and summed. In this case, we form A_{11}, the scalar value 45.

56. **Correct answer: (B)**
    Because of the angle bisection and the presence of right angles, triangles ABC and AED are similar. That is, they have the same set of angles and have sides that are proportional. The hypotenuses AC and AD are corresponding sides, as are the pairs AE, AB and ED, BC. We can therefore set up the proportionality

    $$\frac{AE}{AB} = \frac{AD}{AC}$$

    from which we obtain answer (B).

57. **Correct answer: (C)**
    The transpose A^T of a (square) matrix A is formed by making the substitution $A_{ij} \to A_{ji}$. The inverse of a matrix A is the matrix A^{-1} such that AA^{-1} is the identity matrix I. The condition for orthogonality is therefore $A^T = A^{-1}$, or, multiplying on the left by the original matrix, $AA^T = AA^{-1} = I$. In component form for a 2 by 2 matrix, this becomes

    $$\begin{pmatrix} a & b \\ c & d \end{pmatrix} \begin{pmatrix} a & c \\ b & d \end{pmatrix} = \begin{pmatrix} 1 & 0 \\ 0 & 1 \end{pmatrix}$$

    This gives us four simultaneous equations to solve, three of which are independent:

$$a^2 + b^2 = 1$$
$$c^2 + d^2 = 1$$
$$ac + bd = 0$$

We can either check each matrix in turn for these conditions, or we can note that any matrix of the form

$$\begin{pmatrix} \sin(x) & \cos(x) \\ -\cos(x) & \sin(x) \end{pmatrix}$$

will satisfy them. Matrix (C) is this matrix for the case $x = 45°$. The most direct approach is to multiply each matrix by its transpose to see if the identity matrix obtains.

58. **Correct answer: (C)**
The triangle inequality states

$$|\vec{A} + \vec{B}| \leq |\vec{A}| + |\vec{B}|$$

Writing $\vec{A} - \vec{B}$ as $\vec{A} + (-\vec{B})$, we have

$$|\vec{A} + (-\vec{B})| \leq |\vec{A}| + |-\vec{B}|$$

or

$$|\vec{A} - \vec{B}| \leq |\vec{A}| + |\vec{B}|$$

It is also possible to obtain this answer by trying a few special cases for vectors $\vec{A}$ and $\vec{B}$. The answer is obvious if the triangle inequality is understood graphically.

59. **Correct answer: (C)**
The rank of a matrix is the number of linearly independent rows or columns it has. Here, rows 2 and 3 are linearly dependent so the rank cannot be 3. (Equivalently, the matrix has a zero determinant so its rows are linearly related.) Rows 1 and 2 are linearly independent so the rank is 2. Alternately, the square submatrix

$$\begin{pmatrix} 1 & 2 \\ 1 & 1 \end{pmatrix}$$

is not singular, so the rank must be 2.

60. **Correct answer: (A)**
The indefinite integral of $(1 + 1/x)$ is $(x + \ln(x))$. Evaluating this

from 1 to e, we have

$$(e + \ln(e)) - (1 + \ln(1))$$

$\ln(1)$ is 0, and $\ln(e)$ is 1 so the integral is

$$(e + 1) - (1) = e$$

61. Correct answer: **(A)**

$$\cos(2x) = \cos^2(x) - \sin^2(x)$$

so that the integrand reduces to

$$\ln[\sin^2(x) + \cos^2(x)] = \ln(1) = 0$$

If the integrand is identically zero, the integral is zero as well.

62. Correct answer: **(B)**
The two most straightforward ways to do this integral are to make the substitution

$$\tfrac{1}{2}\sin(2x) = \sin(x)\cos(x)$$

and integrate the sin function or to recognize that $\sin(x)\cos(x)\,dx$ is of the form $y\,dy$ if $y = \sin(x)$, whence the integral is $y^2/2$. Proceeding with the first of these, the integral becomes

$$\frac{1}{2}\int_0^{\pi/4} \sin(2x)\,dx$$

Letting $u = 2x$, $du = 2dx$ and including the change of integration limits, we have

$$\frac{1}{4}\int_0^{\pi/2} \sin(u)\,du$$

The indefinite integral is $-\tfrac{1}{4}\cos(u)$, which evaluates to $-\tfrac{1}{4}[0 - 1] = \tfrac{1}{4}$.

63. Correct answer: **(A)**

$$\int_a^b g'(y)\,dy$$

is $g(b) - g(a)$. It doesn't matter that a and b in this case are functions of a different variable x. The integral is

$$g\left(\frac{1}{x}\right) - g(1)$$

64. Correct answer: **(C)**
    The indefinite integral of $\cos(x)$ is $\sin(x)$ [not $-\sin(x)$]. Evaluating this at the upper limit, we obtain

    $$\sin\left(\frac{5\pi}{2}\right) = \sin\left(\frac{\pi}{2}\right) = 1$$

    Evaluating this at the lower limit, we obtain $\sin(0) = 1$. The integral is therefore 1.

PART C

65. Correct answer: **(A)**
All straight lines other than vertical ones (x = constant) cross the y axis. The first equation may be written as $y = 2(x + 1)$ so the intersection occurs at $y = 2$. The second equation may be written as $y = 4/x$. As $x = 0$ is approached from above or below, the value of y diverges to infinity and minus infinity, respectively, so no intersection occurs. Finally, the third equation defines a circle of radius 2 whose center is at $(4, -4)$; clearly no intersection occurs.

66. Correct answer: **(A)**
A quick sketch reveals no intersection. The straight line passes through quadrants 1, 2, and 3, and the circle lies in quadrant 4. An alternative to a sketch is an algebraic substitution of equation 1 into equation 3. From the former,

$$y = 2(x + 1)$$

and upon insertion into the latter we obtain

$$5x^2 + 16x + 48 = 0$$

The solution to the quadratic equation

$$ax^2 + bx + c = 0$$

is

$$y = \frac{-b \pm \sqrt{b^2 - 4ac}}{2a}$$

Here we find $b^2 - 4ac$ is negative, so there is no real solution and therefore no intersection.

67. Correct answer: **(B)**
$g(X)$ is the area under the curve f from $x = 0$ up to $x = X$. This is clearly zero for the case where $X = 0$ since the integration limits are equal.

68. Correct answer: **(E)**
Recalling that $g(X)$ is the area under the curve, we estimate the negative area from $x = 0$ to $x = 1.5$ as about 1.5 square units (a

triangular shape, so $A = bh/2$) and the positive area from $x = 1.5$ to $x = 7.5$ as 12 square units. The total area is then about 10.5 square units.

69. Correct answer: **(C)**
$g(x)$ is the integral of $f(x)$. If $f(x)$ remains a negative constant $-k^2$ for all large x, the integral asymptotes to $-k^2 x$ for sufficiently large x.

70. Correct answer: **(A)**
$g(X)$ is positive whenever the total area from $x = 0$ to $x = X$ is positive. g first becomes positive at about 3 when the two triangular-shaped regions have equal and opposite areas. g remains positive until the large positive region ending at $x = 7.5$ is canceled out by the flat negative portion that starts at the same point. This happens at about $x = 11$.

71. Correct answer: **(B)**
The integral of a curve increases as long as the function is positive, that is, the contribution to the area under the curve is positive. The function $f(x)$ is positive from 1.5 to 7.5.

72. Correct answer: **(E)**
Similarly to Problem 71, the integral of a curve is a decreasing function of x as long as the contribution to the integral is negative, that is, the function is negative. This occurs in (0, 1.5) and (7.5, 12).

73. Correct answer: **(E)**
When one object is above the other the distance between them is a minimum. There are five such minima in the figure.

74. Correct answer: **(E)**
The connecting line can only intersect the center of the cube if A and B are at diagonally opposite ends of the cube. Suppose that A is initially on top of B at a vertex, then top-down views of their connecting line at later times would show parallel lines at a 45° angle to an edge (except when the line is vertical). These lines hit the center of the cube twice per complete rotation around the cube. Similarly, if A is on top of B in the middle of an edge, initially the connecting line is parallel to a face at later times, and the center of the cube is still hit twice per rotation. This is true in general and since the figure of Problem 73 shows two complete rotations, the center is hit four times.

75. Correct answer: **(B)**
    The distance between A and B is a minimum when one is on top of the other; their separation is then the side of the cube. From the figure, the minimum distance is 4 units.

76. Correct answer: **(B)**
    An equivalent question is, How often does the slope of this curve change from positive to negative? This happens at $t = 2.5$ sec, 7.5 sec, and so on, for a total of four times. Alternately, considering the case where A is initially above B on a corner, a few sketches reveal that twice each complete rotation around the cube the objects change their behavior from approaching to receding.

77. Correct answer: **(C)**
    This is subtle. We have determined that the side of the cube is 4 units. At $t = 2.5$ sec the separation is 5.6 units. The objects could not be at diagonally opposite corners of the cube or their separation would be $\sqrt{3}$ times as large as a side. It appears that A and B could be at opposite edges of a single face because their separation is $\sqrt{2}$ times a side; however, they must be farther apart than one edge if they are to be at a new minimum at $t = 5$ sec. (Also they would spend the next 2.5 sec at a constant separation!) Objects A and B must be at the center of opposite edges at $t = 2.5$ sec because their separation is $\sqrt{2}$ times a side *and* they will be at a new minimum at $t = 5$ sec.

78. Correct answer: **(C)**
    The antisymmetric extension of the figure appears below. From the symmetry, answers (A) and (B) are ruled out. Answer (D) is not possible since it has no symmetry with respect to the origin—also we recognize (D) as a shifted parabola (not enough wiggles). Answer (E) has the appropriate symmetry, but the shape of the function is incorrect. Answer (C) has the appropriate symmetry and, as a cubic, enough wiggles near the origin.

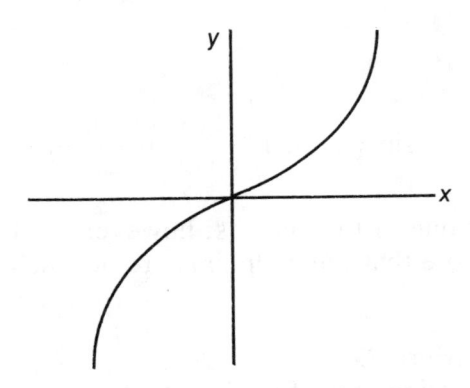

79. **Correct answer: (B)**
    The relationship between the two temperature scales is given by
    $$Q = 2P - 200$$
    Thus, when $P = 50°$, $Q = -100°$.

80. **Correct answer: (B)**
    The straight line equation is rewritten as
    $$y = \frac{2}{3}x + \frac{5}{6}$$
    Since the slope is 2/3, all perpendicular lines have a slope of $-3/2$ so that the product of the slopes is 1. This property arises from the equations
    $$\tan(\theta) = \text{slope of first line}$$
    $$\tan(90 + \theta) = \text{slope of perpendicular line}$$
    The latter expands to
    $$\frac{-\cos(\theta)}{\sin(\theta)} = \frac{-1}{\text{slope of first line}}$$

81. **Correct answer: (D)**
    The function is easily simplified through the definitions
    $$\tan(x) = \frac{\sin(x)}{\cos(x)}$$
    and
    $$\csc(x) = \frac{1}{\sin(x)}$$
    We then have
    $$\sin(x)\cos(x)\left(\frac{\sin(x)}{\cos(x)}\right)\left(\frac{1}{\sin(x)}\right) = \sin(x)$$
    This is not one of the choices; however, in $(0, \pi)$ $\sin(x)$ is always positive. Note that the endpoints are not included in the interval.

82. **Correct answer: (D)**
    Direct evaluation of the function at $x = 0$ gives rise to the indetermi-

nate form 0/0; therefore we may use l'Hôpital's Rule, which states that the

$$\text{limit}_{x \to x_0} \frac{f(x)}{g(x)} = \text{limit}_{x \to x_0} \frac{f'(x)}{g'(x)}$$

Here $f'(x) = e^x - 2e^{2x}$ and $g'(x) = -\sin(x)$. Evaluation of f'/g' at $x = 0^+$ indicates a divergence to positive infinity.

83. **Correct answer: (E)**
The energy is emitted uniformly over a hemisphere whose surface area is $2\pi r^2$. Thus

$$I = \frac{E_{\text{total}}}{2\pi r^2}$$

Since no energy leaks out the sides of this growing hemisphere, we may equate the total energies for any values of r. Thus, $I_1 r_1^2 = I_2 r_2^2$. Since the ratio of r_1 to r_2 is $3:1$, the ratio of I_1 to I_2 is $1:9$. Therefore,

$$I(15\ m) = \frac{5}{9}\ \text{W/m}^2$$

84. **Correct answer: (C)**
To calculate an implicit derivative, we compute the total differential of an equation, group dy and dx terms and solve for dy/dx:

$$d(y^2) = 2y\ dy = d[x \sin(y)] = x \cos(y)\ dy + \sin(y)\ dx$$

Then

$$dy\ [2y - x \cos(y)] = \sin(y)\ dx$$

$$\frac{dy}{dx} = \frac{\sin(y)}{2y - x \cos(y)}$$

85. **Correct answer: (D)**
Because the answer is to be expressed in fundamental SI units, answers (A), (B), and (C) are ruled out. Substituting in the fundamental units for ρ [density] = [mass/volume] = kilogram per cubic meter, ω [frequency] = 1/second, and r [distance] = meters, the expression $\rho \omega^2 r^2$ evaluates to kg/(m * sec²). Alternately, the units of pressure are always the units of force/area = newton per square meter = (kilogram meters per second squared) per square meter.

PART D

86. **Correct answer: (D)**

    Classically, the oscillating electric field associated with the incident light causes the electrons bound to the surface to oscillate, absorbing energy until they overcome their binding and leave the surface. The photocurrent is therefore proportional to the incident light intensity, the time of exposure, and the size of the incident beam and is independent of the frequency of the incident wave. Quantum mechanically (and experimentally!) we find that the photocurrent appears almost immediately (before an oscillating electron should have overcome electron binding), that there is a cutoff frequency below which no electrons are ejected, and as in the classical picture the photocurrent is proportional to the incident intensity. In the quantum mechanical view, the incident beam consists of photons of a certain frequency and associated energy, each of which may cause the release of a single electron.

87. **Correct answer: (B)**

    Newton's Second Law, $F = ma$, may be written in its impact-momentum form as $\bar{F}\Delta t = m\Delta v$, where $\bar{F}$ is the time-averaged force. Here Δt is x seconds, Δv is the change in velocity of the first ball, $-v$, and so

    $$\bar{F} = \frac{-mv}{x}$$

88. **Correct answer: (C)**

    $$V_{rms} = \sqrt{\frac{1}{T}\int_0^T v(t)\,dt}$$

    Equations of line

    $$ab: v(t) = 2$$
    $$bc: v(t) = -2(t-2)$$

    Hence, $V^2(t)$ becomes

    $$V_{rms} = \sqrt{\frac{1}{2}\int_0^1 4\,dt + \frac{1}{2}\int_1^2 4(t^2 - 4t + 4)\,dt}$$

    $$V_{rms} = \sqrt{\frac{8}{3}} = 2\sqrt{\frac{2}{3}}\,V$$

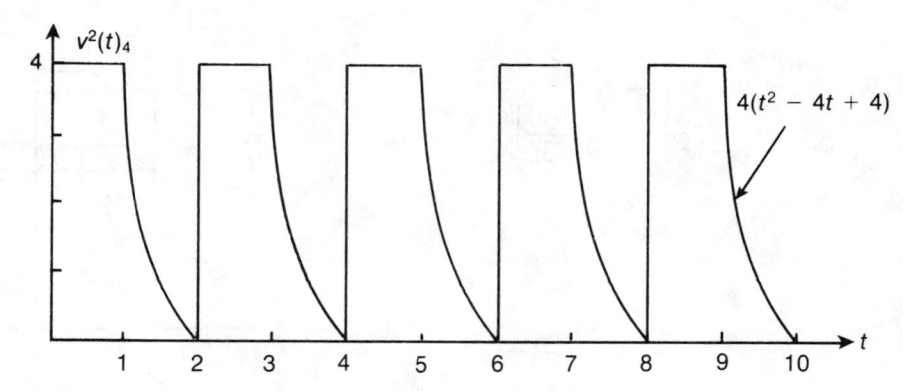

[Graph showing $v^2(t)$ with peaks of 4 at integer times, decaying curves following $4(t^2 - 4t + 4)$, with t-axis marked 1 through 10]

89. **Correct answer: (B)**

    From the principles of dynamic similitude, the Reynolds number in the prototype will have to be equal to the Reynolds number in the model

    $$\frac{v_P D_P}{\nu_P} = \frac{v_m D_m}{\nu_m}$$

    where,

    v = velocity
    D = characteristic length
    ν = kinematic viscosity
    P = prototype
    m = model

    Since both fluids are the same, $\nu_P = \nu_m$

    $$v_m = v_p \frac{D_P}{D_m} = 4\left(\frac{10}{1}\right) = 40 \text{ ft/sec}$$

90. **Correct answer: (B)**

    The equation for the moment of inertia of the area (M) about an axis AA is

    $$M = \int y^2 \, dA$$

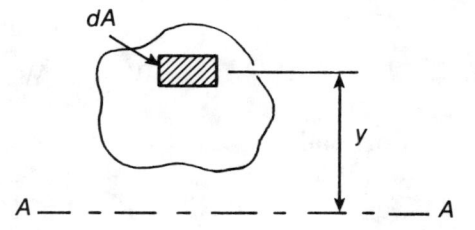

    where dA is an area segment and y is the distance through the centroid of the segment to the axis in question. For the given area,

$$dA = bdy$$

$$M = \int_0^h y^2 b\,dy = \frac{by^3}{3}\Big]_0^h = \frac{bh^3}{3}$$

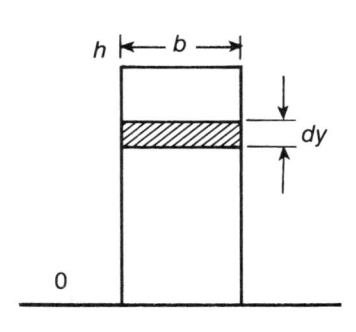

91. Correct answer: **(D)**

   Pressure at bottom = pressure air + pressure water

   Pressure air = 20 psi given

   Pressure water = γh = 62.5 lb/ft³ × 10 ft

   = 625 lb/ft²

   $$\frac{625}{144}\frac{\text{lb/ft}^2}{\text{in.}^2/\text{ft}^2} = 4.34 \text{ psi}$$

   Therefore, the answer is

   Gauge pressure at bottom = 20 + 4.34 = 24.34 psi

   To find the absolute pressure, the atmospheric pressure must be added to the gauge pressure.

92. Correct answer: **(C)**
    Use the power triangle

    $$W = P + jQ$$

    For load A

    $P = 20$ W

    $\theta = \cos^{-1}\theta \cong 37°$

    $Q = -P\tan\theta = -15$ VAR

    $W = 25$ W

    For load B

    $P = 16$ W

    $\theta = \cos^{-1}\theta \cong 53$

    $Q = -21.33$ VAR

    $W = 26.67$ W

Total real power consumed in the network is

$$P_T = 20 + 16 + \frac{120^2}{R}$$

Reactive power is

$$Q_T = -15 - 21.33 = -36.33 \text{ VAR}$$

Overall power factor

$$\cos \theta = 0.9$$

or

$$\theta = 25.84°$$

$$\tan \theta = \frac{Q}{P}$$

$$\tan 25.84 = \frac{36.33}{36 + (120^2/R)}$$

$$R = 368.75 \text{ }\Omega$$

93. **Correct answer: (B)**
    The fins of the car radiator essentially provide more surface area for heat transfer, thereby enhancing the rate of heat transfer.

94. **Correct answer: (D)**
    The period of a pendulum depends only on the local value of g and the length of the string r. We derive this result from Newton's Second Law of Motion for rotation, $\vec{\tau} = I\vec{\alpha}$, where $\vec{\tau}$ is the torque, I is the moment of inertia of a point mass $= mr^2$, and $\vec{\alpha}$ is the resulting angular acceleration. The torque is $\vec{r} \times \vec{F}$, where $\vec{F}$ is $m\vec{g}$ and the cross product introduces a factor of $\sin(\theta)$, where θ is the angular deflection of the string. Putting this all together and making the small angle approximation $\sin(\theta)$ is approximately θ for small θ (in radians), we have $\ddot{\theta} = -(g/r)\theta$, which has sinusoidal solutions of the form $\theta = \theta_0 \sin(\omega t)$ for $\omega = \sqrt{g/r}$. We then use $\omega = 2\pi f = 2\pi/T$, where T is the period of the motion. For very small amplitude motion, the period of the motion is independent of the amplitude.

95. **Correct answer: (C)**
    From the definition of entropy, s,

$$ds = \left(\frac{\delta Q}{T}\right)_{rev}$$

where Q is the heat flow and T the absolute temperature, it is evident that the entropy remains constant in a reversible adiabatic ($\delta Q = 0$) process.

96. Correct answer: **(B)**
A unit cell consists of the body-centered atom and the pieces of the eight atoms that lie on the boundary of the cell. Each of these atoms is equally shared by eight unit cells so the number of atoms owned by a single cell is $1 + 8(1/8)$, or 2.

97. Correct answer: **(C)**
Even without a clear understanding of emittance and absorptivity it should be clear that if absorptivity/emittance is larger for one body than another, under identical conditions the former has a stronger tendency to retain thermal energy.

98. Correct answer: **(B)**
The forces in the stretched line are the weight, mg, of the bird vertically downward and, in each of the two halves of the line, a tension T directed along the line. The upward components of the tension must equal the weight and the tension must be at least 200 lb if the line breaks. If the angular deflection is θ from the horizontal, then the total upward force is $2T * \sin(\theta)$. Thus

$$mg = 1/4 = 2(200) \sin(\theta)$$

or

$$\sin(\theta) = \frac{1}{1600}$$

The angle is clearly small, so we may make the approximation

$$\sin(\theta) = \theta$$

where θ is in radians.

$$\theta = \frac{1}{1600} \text{ rad} = \frac{1}{1600} (57.3°) = 0.03°$$

99. Correct answer: **(B)**
The formula $\Delta q = mc\Delta T$ tells us how much energy Δq is required to bring a mass m of substance with a specific heat c through a temperature difference ΔT provided there is no phase change. This equation may be used to figure out the energy required to bring the liquid

to its boiling point and the gas from the boiling temperature to its final temperature, but to this we must add the energy involved in the phase change, the heat of vaporization times the mass of liquid.

100. Correct answer: **(C)**
Each molecule of SO contains one sulfur atom (atomic weight 32) and one oxygen atom (atomic weight 16), so the original 10 kg of material contains 6.67 kg of sulfur and 3.33 kg of oxygen. After complete combustion, each sulfur atom combines with two oxygen atoms so we have 6.67 kg of sulfur and 6.67 kg of oxygen. Because we were asked for the weight rather than the mass, we multiply by g, which is about 10 m/sec² and obtain 133 N.

101. Correct answer: **(A)**
Experience tells us that the elevator must be accelerating if we are to feel a weight change, and the direction must be upward. From Newton's Second Law, one's effective weight is $m(g + a)$. If the change in the weight is 20%, then a must be 20% of g, or about 2 m/sec².

102. Correct answer: **(E)**
Sum flow into joint equals flow out of joint

$$Q_A + Q_B = Q_C$$

$$Q_A = A_A V_A = \frac{\pi(2)^2}{4}(2) = 6.28 \text{ cfs}$$

$$Q_B = A_B V_B = \frac{\pi(1)^2}{4}(4) = 3.14 \text{ cfs}$$

Therefore,

$$Q_c = 6.28 + 3.14 = 9.42 \text{ cfs}$$

103. Correct answer: **(E)**
Gibbs phase rule states that $F = C - P + 2$, where F is the number of degrees of freedom (the number of variables that may be adjusted independently), C is the number of components, and P is the number of phases present. Here $F = 2 - 4 + 2 = 0$, so there are no adjustable parameters. Any change in $T, P,$ or V will result in one or more phases disappearing.

104. Correct answer: **(C)**
This problem is solved by the process of elimination if you do not remember that the velocity of an object in uniform circular motion

may be written as

$$\bar{v}(t) = \sin(\omega t)\hat{x} + \cos(\omega t)\hat{y}$$

The height of a projectile is a quadratic in time, not a sinusoid. Similarly, the vertical component of a projectile's velocity is a linear function of time. If an object experiences a linearly growing force, then solving Newton's Second Law we again find a quadratic solution.

105. Correct answer: **(B)**
The radiation emitted by the interior walls of a cavity is considered isotropic and regarded as blackbody radiation. The radiation emitted by a blackbody is a function of the temperature only, given by the Stefan–Boltzmann equation

$$E_b = \sigma T^4$$

where E_b is the emissive power or radiant emittance of a blackbody, σ is the Stefan–Boltzmann constant, and T is the absolute temperature.

106. Correct answer: **(D)**
The interest payment for the first year is

$$0.11 \times 120{,}000 = \$13{,}200$$

The capital cost is reduced by the difference between the total payment ($14,000) and the interest payment ($13,200). First year's reduction of the capital cost = $14,000 − $13,200 = $800.

107. Correct answer: **(C)**
The power dissipated in a resistor R subject to the potential V is V^2/R. Here V is a constant and R presumably has the temperature dependence $R = R_0(1 + \varepsilon T)$. Answers are in terms of time dependence and not temperature dependence so we need to estimate the time dependence of the temperature of the resistor. We certainly expect the resistor to heat up, so we try

$$T = T_0(1 + ft)$$

Combining these equations for R and T in $V^2 R$ we obtain answer (C).

108. Correct answer: **(C)**
A 20-kg block has a 200-N normal force, half of which is experienced by each surface. The maximum frictional force is μN, where

μ is the coefficient of friction and N is the normal force. The total maximum horizontal frictional force is therefore $100N * 0.1 + 100N * 0.2 = 30$ N. If the applied force has the magnitude $(10 + 10t)N$, a force of 30 N occurs at $t = 2$ sec.

109. **Correct answer: (D)**
 For all media

$$\alpha + \rho + \tau = 1$$

 Most solids encountered in engineering practice are opaque to thermal radiation, resulting in $\tau = 0$. Thus $\alpha + \rho \simeq 1$ for most solids.
 Many liquids are transparent to thermal radiation, that is, $\alpha = 0$. The $\rho + \tau \simeq 1$ for many liquids.
 Most gases have high values of τ and low values of α and ρ. For all practical purposes, for most gases, $\alpha = \rho = 0$ and $\tau \simeq 1$.

110. **Correct answer: (D)**
 The radiation incident upon the body is characterized by the temperature of the oven walls, and the absorptivity of the body for this radiation is 0.5. Thus the rate of energy absorption G is

$$G = \alpha \sigma T^4$$
$$= 0.5 \times 0.1714 \times 10^{-8} \frac{\text{Btu}}{\text{hr-ft}^2\text{-}R^4} \times (2000 + 460)^4 R^4$$
$$= 3.16 \times 10^5 \frac{\text{Btu}}{\text{hr-ft}^2}$$

111. **Correct answer: (D)**

$$\sum \text{Force vertical direction} = ma = 0$$
$$\text{Buoyant force} = \text{weight of cube}$$
$$\text{Buoyant force} = \text{water displaced by cube}$$
$$= SG \text{ water} \times l \times l \times (h + 0.1l)$$
$$SG \text{ water} = 1.0$$
$$\text{Weight cube} = 0.6 \times l \times l \times l$$

Therefore,
$$0.6l^3 = l^2(h + 0.1l)$$
$$0.6l = h + 0.1l$$

and
$$h = 0.5l$$

190 / Graduate Record Examination in Engineering

112. Correct answer: **(D)**
For the given velocity distribution the flow pattern is

The maximum velocity is at $r = 0$. Therefore,

$$v = C(r_0^2 - 0) = Cr_0^2$$

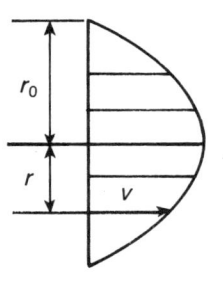

113. Correct answer: **(E)**
Although conservation of kinetic energy is not always satisfied in two-particle collisions, conservation of linear and angular momentum is satisfied. The initial angular momentum of the system is not zero even though the sole moving object is in pure translation. $\vec{L} = \vec{r} \times \vec{p}$, $\vec{p}$ is not zero because there is straight line motion, and $\vec{r}$ is not zero for any choice of origin that does not lie along $\vec{p}$. This initial angular momentum is shared with the stick during the collision.

114. Correct answer: **(B)**
The wave nature of light is clearly demonstrated by the diffraction of light through an aperture or a crystal. Similarly, the wave nature of matter is well illustrated by the *diffraction* of massive neutrons through a crystal. Diffraction, interference, and polarization are phenomena associated with waves.

115. Correct answer: **(B)**
The weight of the man on Mars is the force exerted on him by Martian gravity and is determined by

$$F = \frac{ma}{g_c} = \frac{175 \text{ lbm} \cdot 12.27 \text{ ft/sec}^2}{32.174 \dfrac{\text{lbm}}{\text{lbf}} \cdot (\text{ft/sec}^2)}$$

$$= 66.7 \text{ lbf}$$

Note that the man's mass on Mars is the same as that on earth.

116. Correct answer: **(C)**
Although the local growth mechanism for polymers is linear, the molecule as a whole has little motivation to remain linear—it may bend or twist or coil. There is no significant force tending to pull it along a line.

117. **Correct answer: (C)**
Many chemical reactions are temperature dependent as you may recall from attempts to start a car on a cold winter morning. Low concentrations of reactants mean low concentrations of product. A catalyst, by definition, increases the reaction rate, and electromagnetic radiation is a catalyst for many reactions (such reactions are said to be photosensitive). By elimination, we are left with the shape and size of the reaction vessel. This is a reasonable answer because the length scale of the vessel is thousands of times larger than the molecules whose interactions determine the reaction rate.

118. **Correct answer: (B)**
Replacing the base circuit by its Thevenin's equivalent yields the following circuit.

where

$$V = \frac{R_2 V_{cc}}{R_1 + R_2} = \frac{4 \times 15}{80 + 4} = 0.714 \text{ V}$$

$$R_b = \frac{R_1 R_2}{R_1 + R_2} = \frac{4 \times 80}{84} = 3.81 \text{ k}\Omega$$

$$I_c = \beta I_b$$

$$I_e = (1 + \beta) I_b$$

Using *KVL* in the base loop,

$$V = V_{BE} + R_b I_b + R_e I_e$$
$$0.714 = 0.6 + 3.81 I_b + 0.2 \times 51 I_b \rightarrow I_b = 8.317 \text{ }\mu\text{A}$$
$$I_c = 50 \times 8.317 = 406.85 \text{ }\mu\text{A}$$
$$I_e = I_b + I_c = 415.17 \text{ }\mu\text{A}$$

Hence,

$$V_{CE} = V_{CC} - I_C R_C - I_E R_E$$
$$= 15 - 406.85 \times 10^{-6} \times 2 \times 10^3 - 415.17 \times 10^{-6} \times 200$$
$$V_{CE} = 14.1 \text{ V}$$

192 / **Graduate Record Examination in Engineering**

119. Correct answer: **(D)**
    From the law of continuity,

    $$\sum Q_{in} = \sum Q_{out}$$
    $$Q_{in} = 10 \text{ m}^3/\text{sec}$$
    $$Q_{out} = Q_A + \frac{4(\text{m}^3/\text{sec})/\text{m} \times 4 \text{ m}}{2} = QA + 8 \text{ m}^3/\text{sec}$$

    Therefore,

    $$10 = Q_A + 8$$

    and

    $$Q_a = 2 \text{ m}^3/\text{sec}$$

120. Correct answer: **(B)**
    The oscillating magnetic field in the solenoid means that the flux linkage to the loop oscillates at the same frequency as the oscillating magnetic field. As the flux linkage oscillates, the voltage across the loop changes phase just as rapidly according to Faraday's Law, $E = -d\Phi/dt$, where Φ is the magnetic flux. An oscillating voltage in the loop results in an oscillating current. All oscillations are at the same frequency because neither proportionalities nor derivatives change the frequency of a sinusoid.

121. Correct answer: **(D)**
    Consider two stationary men standing near a moving conveyor belt. The first man, the source, throws pies on the belt at a constant rate (frequency), and since the belt moves at constant speed they are a constant distance (wavelength) apart. The second man, the receiver, measures the same frequency and wavelength in this simplest case. It is now possible to analyze source motion, receiver motion, and medium motion in terms of the motion of the men and the belt.
      In the last case, consider the conveyer apparatus to be moving relative to both men. This is clearly the same as the belt just moving faster than before. The receiver must see pies approaching him at the same frequency as before (where would other pies come from? disappear to?), but the wavelength must be larger since each pie is carried a larger distance from the source before the next pie is dropped.

122. Correct answer: **(B)**
    In civil engineering, the hanging cable problem is known to be solved by the hyperbolic cosine function $(e^x + e^{-x})/2$. We can obtain

this answer by elimination. e^x for negative x is a monotone function of x. The cable might look like a piece of a sine curve but not in $(\pi/2, 3\pi/2)$. The remaining trigonometric functions are singular in the intervals specified.

123. Correct answer: **(B)**
Under a clear, cloudless night, thermal radiation is a significant means of heat loss. (The clear sky can often be taken to be at zero absolute temperature.) The water in the tray will experience heat loss by radiation to the clear sky, often more than the net heat gain by convection. The water temperature will continue to drop until thermal equilibrium is reached. Thus, the water can freeze even though the outdoor temperature is way above the freezing point.

124. Correct answer: **(B)**
The equations for streamlines in a three-dimensional Cartesian field are

1. $vdx = udy$
2. $wdx = udz$
3. $wdy = vdz$

Since we are only interested in the xz plane, use equation 2.

$$\frac{dx}{u} = \frac{dz}{w}$$

or

$$\frac{dx}{10x} = \frac{-dz}{20z}$$

and

$$\frac{2dx}{x} = \frac{-dz}{z}$$

Integrating

$2 \ln x = -\ln z + \ln c$

$\ln x^2 + \ln z = \ln c$

or

$\ln x^2 z = \ln c$

and

$x^2 z = C$

194 / Graduate Record Examination in Engineering

125. Correct answer: **(C)**
 The value of a sum of money P compounded annually at an interest rate i at the end of n years is

$$P = C(1 + i)^n$$

where

$$C = \text{initial sum of money}$$

For the given problem, $P = 13{,}400$, $C = 10{,}000$, $n = 6$

$$13{,}400 = 10{,}000(1 + i)^6$$
$$\ln 1.34 = 6 \ln(1 + i)$$
$$\ln(1 + i) = \frac{0.2927}{6} = 0.0488$$
$$1 + i = 1.05$$
$$i = 0.05$$

or

$$5\%$$

126. Correct answer: **(E)**
 Since there are two couples in a plane, take the sum of the moments as follows:

$$\sum M = +(15 \text{ lb})(4 \text{ ft}) - (20 \text{ lb})(3 \text{ ft})$$
$$= 60 \text{ lb/ft} - 60 \text{ lb/ft}$$
$$= 0$$

127. Correct answer: **(B)**
 For steady-state condition, since the external temperatures are maintained constant, the heat flow through material 1 is the same as that through material 2. Using Fourier's Conduction Law, therefore,

$$k_1 \frac{T_1 - T_2}{x_1} = k_2 \frac{T_2 - T_3}{x_2}$$
$$\frac{k_1}{k_2} = \left(\frac{x_1}{x_2}\right)\left(\frac{T_2 - T_3}{T_1 - T_2}\right)$$

since

$$\frac{x_1}{x_2} = 1$$

and

$$\frac{T_2 - T_3}{T_1 - T_2} < 1$$

Therefore,

$$\frac{k_1}{k_2} < 1$$

or

$$k_1 < k_2$$

PART E

128. **Correct answer: (D)**
One approach to this problem is to solve the equation

$$\int_0^c f(x)\, dx = K * c * f(c)$$

for $f(x)$. However, it should be immediately apparent that $f(x)$ will be a power of x because the integral of x^n is proportional to the function multiplied by the independent variable. If we try $f(x) = x^n$ (a multiplicative constant would fall out of the above equation and we are given $f(1) = 1$), we find

$$\frac{c^{(n+1)}}{(n+1)} = Kc^{n+1}$$

or

$$K = \frac{1}{(n+1)}$$

We simply need to determine the order of the original polynomial. We are told $f(2) = 4$ so the order is 2 and $K = 1/3$.

129. **Correct answer: (C)**

$$\int_a^b f(x)\, dx$$

is the area under the curve $f(x)$ from $x = a$ to $x = b$. Similarly, the second integral is the area under the curve from $x = a$ to $x = c$. By sketching these on a number line, we see that their difference is the area under the curve from $x = c$ to $x = b$, whether positive or negative. An alternative approach is to write the integrals in evaluated form as $F(b) - F(a)$ and $F(c) - F(a)$ (where F is the antiderivative of f), subtract, and go back to the unevaluated form.

130. **Correct answer: (E)**
If the limits of integration were the same, we could simply add the integrands. Since we know nothing of function g in the interval (0, 1), we cannot combine the integrals.

131. Correct answer: **(B)**
The derivative could be computed from a defining equation for $\cos(x)$ such as

$$e^{i\theta} = \cos(\theta) + i * \sin(\theta)$$

but this is one of several fundamental derivatives that should be memorized.

132. Correct answer: **(C)**
From the product rule,

$$d(y * e^{xy}) = y * d(e^{xy}) + e^{xy} * dy = d(2) = 0$$

In the first term, $d(e^{xy})$ is evaluated as

$$e^{xy}d(xy) = e^{xy}(xdy + ydx)$$

Dividing out the common factor of e^{xy} (which is not zero), we obtain

$$y(xdy + ydx) + dy = (xy + 1)dy + y^2 dx = 0$$

or

$$\frac{dy}{dx} = \frac{-y^2}{(1 + xy)}$$

133. Correct answer: **(C)**
The product rule of calculus says that

$$\frac{d}{dx}[f(x)g(x)] = f'(x)g(x) + f(x)g'(x)$$

Here

$$f(x) = \ln(x)$$
$$f'(x) = \frac{1}{x}$$
$$g(x) = e^x$$

and

$$g'(x) = g(x) = e^x$$

Plugging into the general form and pulling out the common factor of e^x, we obtain answer (C).

198 / **Graduate Record Examination in Engineering**

134. Correct answer: **(C)**
If a function is decreasing at a point, then its derivative is negative at that point. If a function decreases everywhere, its derivative must always be negative.

135. Correct answer: **(B)**
The quotient rule of calculus says that the derivative of

$$\frac{f(x)}{g(x)}$$

is

$$\frac{[g(x)f'(x) - f(x)g'(x)]}{g(x)^2}$$

Here we replace $f(x)$ with $f(x^2)$ so $f'(x)$ is replaced by

$$\frac{df(y)}{dy}\left(\frac{dy}{dx}\right)$$

(with $y = x^2$).

136. Correct answer: **(B)**
The Laurent expansion (Taylor series about the origin) of a function $f(x)$ is given by

$$f(0) + \frac{f'(0)x}{1!} + \frac{f''(0)x^2}{2!} + \cdots$$

Here all derivatives of $f(x)$ are e^x and all evaluate to 1. Therefore, no terms disappear, including the constant term $f(0) = 1$. We are left with

$$\sum_{0}^{\infty} \frac{x^n}{n!}$$

137. Correct answer: **(B)**
Simple substitution of $x = 0$ gives the indeterminate form

$$\frac{0}{0}$$

Application of L'Hôpital's Rule is allowed for this form and for

$$\frac{\infty}{\infty}$$

and states

$$\lim_{x \to x_0} \frac{f(x)}{g(x)} = \lim_{x \to x_0} \frac{f'(x)}{g'(x)}$$

if x_0 is the singular point. A first application gives the new form

$$\frac{e^x - 1}{2x}$$

which is again indeterminate. A second application of L'Hôpital's Rule gives the form

$$\frac{e^x}{2}$$

which may be evaluated at $x = 0$ to give 1/2. Another approach is to Taylor expand e^x as

$$1 + x + \frac{x^2}{2}$$

(for small x), which gives the form

$$\frac{x^2/2}{x^2} = \frac{1}{2}$$

immediately.

138. Correct answer: **(B)**
Short of adding many terms of the series, the only approach to this problem is to recognize the form

$$\sum_{\text{odd } n} (-1)^{n+1} \frac{x^n}{n!}$$

as a Taylor series for the function $\sin(x)$. Replacing x with $\pi/2$, we obtain $\sin(\pi/2)$ or 1.

139. Correct answer: **(E)**
In our sample of five objects, three are identical and two are identi-

cal, resulting in

$$\frac{5!}{3!2!} = 10$$

distinguishable configurations. There are only three possible results of our picking the balls: no red balls are found, one is found, or two are found. We will compute the latter two probabilities using binomial coefficients, and add them.

$$\binom{3}{0} \cdot \binom{2}{2} = \frac{3!}{0!3!} \frac{2!}{2!0!} = 1$$

configuration that results in two red balls, and

$$\binom{3}{1} \cdot \binom{2}{1} = \frac{3!}{1!2!} \frac{2!}{1!1!} = 6$$

configurations that result in one red ball. These account for seven of the ten possibilities, and therefore the probability of at least one red ball is 70%.

140. Correct answer: **(D)**
This is similar to Problem 139. There are 10 distinct arrangements of the balls in A. In B there are

$$\frac{6!}{2!4!} = 15$$

such arrangements. In A the number of arrangements that results in two orange balls is

$$\binom{3}{2}\binom{2}{0} = 3$$

In B the number is

$$\binom{2}{2}\binom{4}{0} = 1$$

Thus, out of $10 \cdot 15$ possible arrangements in both boxes, there are $3 * 1 = 3$ arrangements of four orange balls.

$$\frac{3}{150} = 2\%$$